Optical Microring Resonators

Theory, Techniques, and Applications

SERIES IN OPTICS AND OPTOELECTRONICS

Series Editors: **E Roy Pike**, Kings College, London, UK
Robert G W Brown, University of California, Irvine, USA

Optical Microring Resonators

Theory, Techniques, and Applications

V. Van

University of Alberta
Edmonton, Canada

CRC Press

Taylor & Francis Group

Boca Raton London New York

CRC Press is an imprint of the
Taylor & Francis Group, an **informa** business

A TAYLOR & FRANCIS BOOK

CRC Press
Taylor & Francis Group
6000 Broken Sound Parkway NW, Suite 300
Boca Raton, FL 33487-2742

First issued in paperback 2021

ISBN-13: 978-0-367-78254-2 (pbk)
ISBN-13: 978-1-4665-5124-4 (hbk)

Library of Congress Cataloging-in-Publication Data

Names: Van, Vien, 1971- author.
Title: Optical microring resonators : theory, techniques, and applications / Vien Van.
Other titles: Series in optics and optoelectronics (CRC Press)
Description: Boca Raton, FL : CRC Press, Taylor & Francis Group, [2017] | Series: Series in optics and optoelectronics
Identifiers: LCCN 2016040265| ISBN 9781466551244 (hardback ; alk. paper) | ISBN 1466551240 (hardback ; alk. paper)
Subjects: LCSH: Microresonators (Optoelectronics) | Optoelectronic devices. | Optical wave guides. | Nonlinear optics.
Classification: LCC TK8360.M53 V36 2017 | DDC 621.381/045--dc23
LC record available at https://lccn.loc.gov/2016040265

Visit the Taylor & Francis Web site at
http://www.taylorandfrancis.com

and the CRC Press Web site at
http://www.crcpress.com

Contents

Preface

Since their early development in the 1990s, optical microring resonators have become one of the most important elements in integrated optics technology. These simple but versatile structures have found myriad applications in filters, sensors, lasers, nonlinear optics, optomechanics, and more recently, quantum optics. Indeed, it is difficult to imagine an integrated photonic system today that does not utilize a microring resonator as part of the circuit architecture. Given the diversity and ubiquitousness of their applications, the need arises for a systematic review of the technology that can serve as a reference source for engineers and researchers working in the broader field of integrated photonics.

The objective of this book is to provide a concise treatment of the theory, principles and techniques of microring resonator devices, and their applications. It is intended for graduate students and researchers who wish to familiarize themselves with the technology and acquire sufficient knowledge to enable them to design microring devices for their own applications of interest. Rapid advances in integrated optics and microfabrication technologies have enabled microring devices with increasingly more sophisticated designs and superior performance to be developed. However, the underlying working principles of these devices remain the same for the most part. Thus, while attempts are made to highlight some of the recent advances in microring technology, the main focus of the book is to provide a detailed treatment of the underlying theory and techniques for modeling and designing microring devices so that the reader can readily apply the knowledge to their own applications. Toward this aim, numerous numerical examples are given to help illustrate the application of these techniques, as well as to demonstrate what can theoretically be achieved with these devices. Many of the examples are based on the silicon-on-insulator material system, a choice motivated by the growing prevalence of silicon photonics in integrated optics technology.

The book is divided into five chapters. While a basic familiarity with optics is assumed, a brief review of the concepts essential to the understanding of microring resonators and their applications will be given in the relevant chapters. In particular, Chapter 1 will give a review of the theory of optical waveguides, whispering gallery modes, and coupled waveguide systems, which are the basic elements constituting any microring device. Chapter 2 develops the basic formalisms used to analyze simple microring resonators coupled to one or more waveguides. Chapter 3 is devoted to the analysis and design of coupled microring resonators for filter applications. Nonlinear optics and active photonic applications of microring resonators are the subjects of Chapters 4 and 5, respectively.

This book is born out of my research on microring resonators which began at the University of Waterloo, Ontario, Canada, in the late 1990s and subsequently at the University of Maryland, College Park, in the early 2000s. I have in particular benefited tremendously from the mentorship of Professor Ping-Tong Ho as well as from my colleagues at the Laboratory for Physical Sciences at the University of Maryland. The contributions of my graduate students at the University of Alberta have also helped shape a large part of this book, with special acknowledgment to Ashok Prabhu Masilamani, Alan Tsay, Daniel Bachman, Guangcan Mi, and Siamak Abdollahi. I also acknowledge the assistance of many students in the preparation and editing of the book, with special thanks to Jocelyn Bachman, Guangcan Mi, Yang Ren, and Daniel Bachman for reviewing and proofreading parts of the manuscript.

V. Van
University of Alberta
September 2016

CHAPTER 1

Elements of an Optical Microring Resonator

An optical microring resonator is an integrated optic traveling wave resonator constructed by bending an optical waveguide to form a closed loop, typically of a circular or racetrack shape. Light propagating in the microring waveguide interferes with itself after every trip around the ring. When the roundtrip length is exactly equal to an integer multiple of the guided wavelength, constructive interference of light occurs which gives rise to sharp resonances and large intensity buildup inside the microring. The high wavelength selectivity, strong dispersion, large field enhancement, and high quality factor are important characteristics which make microring resonators extremely versatile and useful for a wide range of applications in optical communication, signal processing, sensing, nonlinear optics and, more recently, quantum optics.

Microring and microdisk resonators were first proposed by Marcatili in 1969 for realizing channel dropping filters based on planar optical waveguides (Marcatili 1969b). Similar traveling wave filters based on microwave striplines had been proposed and studied earlier by Coale (1956). However, it was not until the late 1990s that advances in the microfabrication technology for integrated photonic devices enabled optical microring resonators to be realized with high quality factors. Since then, microring and microdisk resonators have been demonstrated for a wide range of applications in various material systems such as silicon-on-insulator (SOI), III–V semiconductors, glass, and polymers.

Broadly defined, an optical resonator is a structure which confines light in all spatial directions. In a microring resonator, this is achieved in two ways: transversely by the refractive index contrast of the dielectric waveguide used to form the microring and longitudinally by the periodic boundary condition imposed by the ring or racetrack structure. The propagation characteristics of the dielectric waveguide, and especially those of the curved waveguide, thus have a major impact on the characteristics of the microring

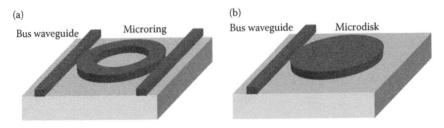

Figure 1.1 Schematic of (a) a microring resonator coupled to two bus (or access) waveguides and (b) a microdisk resonator coupled to one bus waveguide.

resonator. In addition, the manner by which light is coupled into and out of the resonator is also of practical importance. Typically this is achieved via evanescent field coupling between the microring waveguide and one or two external straight waveguides, called access or bus waveguides, by which light is coupled into or out of the resonator. Figure 1.1 illustrates a microring resonator coupled to two bus waveguides and its microdisk variant coupled to a single waveguide.

This chapter provides a brief review of the basic elements constituting a microring resonator, with a view on how their designs and properties may affect the performance of the resonator. Section 1.1 reviews the theory of planar dielectric waveguides and the wave equations governing optical mode propagation. Section 1.2 looks at light propagation in curved waveguides and examines the properties of whispering gallery modes in microdisk and microring resonators. Section 1.3 develops a formalism for analyzing the coupling of waveguide modes in space, which is useful for designing evanescent wave couplers for coupling light into and out of a microring resonator. Finally, Section 1.4 provides an overview of standard fabrication processes for microring devices and highlights several important issues relevant to the practical implementation of these devices.

1.1   Dielectric Optical Waveguides

Dielectric optical waveguides are the basic structures for confining and guiding light in well-defined discrete modes in photonic integrated circuits (PICs). A planar dielectric waveguide consists of a core of refractive index n_1 embedded in other dielectric layers of lower refractive indices. Light is confined within the core due to total internal reflection at the interfaces between the high-index core and the lower-index cladding media. The degree of confinement increases with the refractive index contrast between the core and cladding. For a waveguide with a uniform cladding of refractive index n_2, the index contrast is defined as

$$\Delta n = \frac{n_1^2 - n_2^2}{2n_1^2} \approx \frac{n_1 - n_2}{n_1}. \tag{1.1}$$

The approximation in the above formula is good for low-index contrast waveguides. Some common waveguide material systems and their index contrasts are listed in Table 1.1. Typically, the index contrast ranges from about 1% for weakly confined waveguides based on doped silica materials, to over 40% for strongly confined semiconductor waveguides. In general, high-index contrast (or high-Δn) waveguides are desirable for the miniaturization of PICs since they have smaller dimensions and provide stronger confinement of light, which enables sharp waveguide bends to be realized with low bending loss. On the other hand, polarization-dependent effects and scattering loss also tend to be more pronounced in high-Δn waveguides.

Two basic optical waveguide structures are shown in Figure 1.2: the rib (or ridge) waveguide and the rectangular strip waveguide. From the fabrication point of view, these two structures differ only by the etch depth in defining the waveguide core: in a rib waveguide, the core is etched only to a depth h leaving a residual high-index

Table 1.1   Refractive Indices of Some Common Integrated Optic Waveguide Materials

Core Material	Refractive Index at $\lambda = 1.55$ μm	Index Contrast[a] Δn (%)
Doped silica	1.45–1.5	0.7–4
Polymers	1.45–1.7	0.7–14
SiO_xN_y	1.45–2.0	0.7–24
SiN_x	2.0–2.3	24–30
III–V (InP, GaAs)	3.16, 3.4	40, 41
Si	3.47	41

[a] Index contrast assuming SiO_2 cladding with refractive index $n_2 = 1.44$.

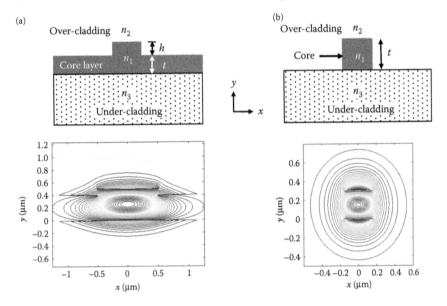

Figure 1.2   Schematic and E_y-field distribution of the quasi-TM mode of (a) a rib (or ridge) waveguide and (b) a rectangular strip waveguide.

layer of thickness t, whereas in a strip waveguide, the core layer is completely etched through. However, the modal characteristics of the two waveguides are quite different, as shown in Figure 1.2. In the rib waveguide, the residual high-index layer causes the mode to expand laterally. Due to the weak lateral confinement, rib wave-guides tend to suffer from larger bending loss than strip wave-guides. As a result, microring resonators are typically designed using strip waveguides to minimize radiation loss due to bending.

Since the performance of a PIC (photonic integrated circuit) depends critically on the properties of the waveguide modes, it is important to obtain a detailed analysis of the propagation charac-teristics of light in the optical waveguide. In the next section, we will derive the wave equation and its various approximations for describing electromagnetic wave propagation in a planar dielectric waveguide. A review of the main methods for solving these equations will also be given.

1.1.1   The vectorial wave equations

The field distributions of a waveguide mode and its associated propagation constant are determined by solving an eigenvalue problem formulated in terms of either the transverse electric (TE) field or transverse magnetic (TM) field. We consider a dielectric waveguide oriented along the z-axis and characterized by a transverse index profile $n(x, y)$. Under the assumption that the transverse index profile is invariant along the propagation axis (the z-axis), the electric and magnetic field distributions of a waveguide mode can be expressed as

$$\mathcal{E}(x,y,z) = \mathbf{E}(x,y)e^{-j\beta z},\tag{1.2}$$

$$\mathcal{H}(x,y,z) = \mathbf{H}(x,y)e^{-j\beta z},\tag{1.3}$$

where β is the propagation constant of the waveguide mode and the time dependence $e^{j\omega t}$ is assumed and suppressed. For the purpose of modal analysis, it is convenient to further decompose the fields $\mathbf{E}$ and $\mathbf{H}$ into a transverse component and a longitudinal component as follows:

$$\mathbf{E}(x,y) = \mathbf{E}_t(x,y) + E_z(x,y)\hat{\mathbf{z}},\tag{1.4}$$

$$\mathbf{H}(x,y) = \mathbf{H}_t(x,y) + H_z(x,y)\hat{\mathbf{z}}.\tag{1.5}$$

The electric and magnetic fields $\mathcal{E}$ and $\mathcal{H}$ satisfy Maxwell's equations,

$$\nabla \times \mathcal{E} = -j\omega\mu_0\mathcal{H},\tag{1.6}$$

$$\nabla \times \mathcal{H} = j\omega\varepsilon_0 n^2(x,y)\mathcal{E},\tag{1.7}$$

where ε_0 and μ_0 are the electric permittivity and magnetic permeability, respectively, of vacuum. By taking the curl of Equation 1.6 and using Equation 1.7 to eliminate $\nabla \times \mathcal{H}$ from the resulting equation, we get

$$\nabla \times \nabla \times \mathcal{E} = n^2 k^2 \mathcal{E},\tag{1.8}$$

where $k = \omega/c$. With the help of the vector identity $\nabla \times \nabla \times \mathcal{E} = \nabla(\nabla \cdot \mathcal{E}) - \nabla^2\mathcal{E}$, we can write Equation 1.8 as

$$\nabla^2\mathcal{E} + n^2 k^2 \mathcal{E} = \nabla(\nabla \cdot \mathcal{E}).\tag{1.9}$$

Substituting $\mathcal{E}(x,y,z) = \mathbf{E}(x,y)e^{-j\beta z}$ into the above equation and making use of the field decomposition in Equation 1.4, we obtain

$$\nabla_t^2 \mathbf{E}_t + (n^2 k^2 - \beta^2)\mathbf{E}_t = \nabla_t(\nabla_t \cdot \mathbf{E}_t) + \nabla_t\left(\frac{\partial E_z}{\partial z}\right), \tag{1.10}$$

where $\nabla_t = \hat{\mathbf{x}}(\partial/\partial x) + \hat{\mathbf{y}}(\partial/\partial y)$. In the absence of free charge, Gauss's law gives

$$\nabla \cdot (n^2 \mathbf{E}) = \nabla_t \cdot (n^2 \mathbf{E}_t) + n^2 \frac{\partial E_z}{\partial z} = 0, \tag{1.11}$$

from which we get

$$\frac{\partial E_z}{\partial z} = -\frac{1}{n^2}\nabla_t \cdot (n^2 \mathbf{E}_t). \tag{1.12}$$

Upon substituting the above expression into Equation 1.10, we obtain the vectorial wave equation in terms of the transverse electric field,

$$\nabla_t^2 \mathbf{E}_t + (n^2 k^2 - \beta^2)\mathbf{E}_t = \nabla_t(\nabla_t \cdot \mathbf{E}_t) - \nabla_t\left[\frac{1}{n^2}\nabla_t \cdot (n^2 \mathbf{E}_t)\right]. \tag{1.13}$$

Equation 1.13 is an eigenvalue problem whose solution gives the transverse field distribution $\mathbf{E}_t$ of an optical mode and its propagation constant β. The effective index of the waveguide mode is defined as $n_{\mathrm{eff}} = \beta/k$.

The terms on the right-hand side of Equation 1.13 account for the polarization coupling between the transverse field components E_x and E_y. Thus, in general, the mode of an optical waveguide is hybrid or vectorial in nature, that is, it contains both E_x and E_y components of the electric field. We can write Equation 1.13 in the form of an eigenvalue matrix equation as (Xu et al. 1994)

$$\begin{bmatrix} P_{xx} & P_{xy} \\ P_{yx} & P_{yy} \end{bmatrix}\begin{bmatrix} E_x \\ E_y \end{bmatrix} = \beta^2 \begin{bmatrix} E_x \\ E_y \end{bmatrix}, \tag{1.14}$$

where the operators in the matrix are given by

$$P_{xx}E_x = \frac{\partial}{\partial x}\left[\frac{1}{n^2}\frac{\partial}{\partial x}(n^2 E_x)\right] + \frac{\partial^2 E_x}{\partial y^2} + n^2 k^2 E_x, \tag{1.15}$$

$$P_{yy}E_y = \frac{\partial^2 E_y}{\partial x^2} + \frac{\partial}{\partial y}\left[\frac{1}{n^2}\frac{\partial}{\partial y}(n^2 E_y)\right] + n^2 k^2 E_y, \tag{1.16}$$

$$P_{xy}E_y = \frac{\partial}{\partial x}\left[\frac{1}{n^2}\frac{\partial}{\partial y}\left(n^2 E_y\right)\right] - \frac{\partial^2 E_y}{\partial y \partial x}, \tag{1.17}$$

$$P_{yx}E_x = \frac{\partial}{\partial y}\left[\frac{1}{n^2}\frac{\partial}{\partial x}\left(n^2 E_x\right)\right] - \frac{\partial^2 E_x}{\partial x \partial y}. \tag{1.18}$$

It is apparent from Equation 1.14 that the operators P_{xy} and P_{yx} give rise to polarization coupling effects. For rectangular waveguides with low to moderate index contrasts, the two lowest-order modes are predominantly linearly polarized along either the principal x- or y-axis. It is often a good approximation to neglect the minor field component of each mode and consider the mode to be either quasi-TE with major field component E_x, or quasi-TM with major field component E_y. Under this semi-vectorial approximation, the cross-polarization coupling terms in Equation 1.14 are neglected so that the equations governing the major field components become

$$P_{xx}E_x = \frac{\partial}{\partial x}\left[\frac{1}{n^2}\frac{\partial}{\partial x}\left(n^2 E_x\right)\right] + \frac{\partial^2 E_x}{\partial y^2} + n^2 k^2 E_x = \beta_{TE}^2 E_x, \quad \text{(quasi-TE)}$$

$$\tag{1.19}$$

$$P_{yy}E_y = \frac{\partial^2 E_y}{\partial x^2} + \frac{\partial}{\partial y}\left[\frac{1}{n^2}\frac{\partial}{\partial y}\left(n^2 E_y\right)\right] + n^2 k^2 E_y = \beta_{TM}^2 E_y. \quad \text{(quasi-TM)}$$

$$\tag{1.20}$$

For low-index contrast waveguides, one may further neglect the spatial index variation in the square bracket terms in the above equations. Under this approximation, the TE and TM modes become identical and are described by the scalar wave equation

$$P_{xx}E = P_{yy}E = \frac{\partial^2 E}{\partial x^2} + \frac{\partial^2 E}{\partial y^2} + n^2 k^2 E = \beta^2 E. \tag{1.21}$$

Figure 1.3 shows the electric field distributions of the two lowest-order modes in an SOI strip waveguide consisting of a Si core of 250 nm thickness and 400 nm width embedded in a SiO_2 cladding. Both the semi-vectorial and full-vectorial solutions of the modes are shown for comparison. Also shown are the effective indices of the modes at the 1.55 μm wavelength. We see that the semi-vectorial and full-vectorial solutions give similar field distributions

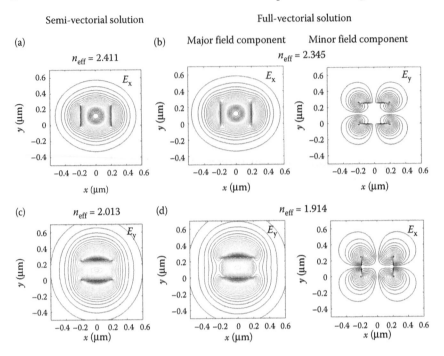

Figure 1.3 Semi-vectorial and full-vectorial solutions of the two lowest-order modes at 1.55 μm wavelength of an SOI strip waveguide (Si core of dimensions 400×250 nm^2 embedded in a SiO$_2$ cladding): (a, b) TE mode, (c, d) TM mode.

for the major field component of each mode, although there is a difference of about 3–5% between the effective index values.

We can also formulate the wave equation in terms of the transverse magnetic field $\mathbf{H}_t$. One advantage of solving for the optical mode in terms of the magnetic field is that the fields H_x and H_y are continuous across all dielectric boundaries. By taking the curl of Equation 1.7 and using Equation 1.6 to eliminate $\nabla \times \mathcal{E}$, we get

$$\nabla^2 \mathcal{H} + n^2 k^2 \mathcal{H} = -\frac{1}{n^2} \nabla n^2 \times (\nabla \times \mathcal{H}). \tag{1.22}$$

Substituting Equations 1.3 and 1.5 into the above equation, we obtain the following vectorial wave equation in terms of the transverse magnetic field,

$$\nabla_t^2 \mathbf{H}_t + (n^2 k^2 - \beta^2)\mathbf{H}_t = -\frac{1}{n^2}(\nabla_t n^2) \times (\nabla_t \times \mathbf{H}_t). \tag{1.23}$$

The above equation can be written explicitly in a component form as (Xu et al. 1994)

$$
\begin{bmatrix} Q_{xx} & Q_{xy} \\ Q_{yx} & Q_{yy} \end{bmatrix} \begin{bmatrix} H_x \\ H_y \end{bmatrix} = \beta^2 \begin{bmatrix} H_x \\ H_y \end{bmatrix},
\tag{1.24}
$$

where

$$
Q_{xx} H_x = \frac{\partial^2 H_x}{\partial x^2} + n^2 \frac{\partial}{\partial y} \left(\frac{1}{n^2} \frac{\partial H_x}{\partial y} \right) + n^2 k^2 H_x,
\tag{1.25}
$$

$$
Q_{yy} H_y = n^2 \frac{\partial}{\partial x} \left(\frac{1}{n^2} \frac{\partial H_y}{\partial x} \right) + \frac{\partial^2 H_y}{\partial y^2} + n^2 k^2 H_y,
\tag{1.26}
$$

$$
Q_{xy} H_y = \frac{\partial^2 H_y}{\partial x \partial y} - n^2 \frac{\partial}{\partial y} \left(\frac{1}{n^2} \frac{\partial H_y}{\partial x} \right),
\tag{1.27}
$$

$$
Q_{yx} H_x = n^2 \frac{\partial}{\partial x} \left(\frac{1}{n^2} \frac{\partial H_x}{\partial y} \right) + \frac{\partial^2 H_x}{\partial y \partial x}.
\tag{1.28}
$$

In general, the vectorial wave equations (1.13) and (1.23) do not have analytical solutions and must be solved numerically. Many efficient numerical techniques have been developed for solving these equations for waveguides with arbitrary cross-sections and index profiles, the most popular ones being the finite difference method (Xu et al. 1994) and the finite element method (Rahman and Davies 1984, Koshiba 1992).[*][†] Approximate methods for computing the effective index are also available, such as Marcatili's method (Marcatili 1969a), the effective index method (EIM) (Knox and Toulios 1970), and perturbation methods (Chiang 1993). These methods generally give good approximations for low index contrast waveguides or for modes far from cutoff. Despite its approximate nature, the EIM has found widespread use in the analysis of planar waveguides, even for high-index contrast waveguides, thanks

---

[*] The finite element method is typically formulated based on either Equation 1.9 for the electric field or Equation 1.22 for the magnetic field.

[†] Commercial software for computing the field distributions and effective indices of optical waveguide modes are also available, such as COMSOL, RSoft, Optiwave, and Lumerical.

to its simplicity and intuitive approach. Given the importance of the method for waveguide analysis, we will briefly review the key aspects of the EIM method below.

1.1.2 The EIM and solutions of the one-dimensional slab waveguide

The basic idea of the EIM method (Knox and Toulios 1970) is the successive approximations of a two-dimensional (2D) rectangular waveguide by one-dimensional (1D) slab waveguides, which can be separately analyzed. The procedure is illustrated in Figure 1.4 for both a rib waveguide and a strip waveguide. In the first approximation, the vertical index profile in each of the core and cladding regions (regions I and II) is replaced by the effective index of the corresponding 1D, y-confined slab waveguides with y-dependent index profiles. This procedure reduces the 2D waveguide to a 1D, x-confined slab waveguide with an x-dependent effective index distribution $n_{eff}(x)$. The effective index of the equivalent slab waveguide

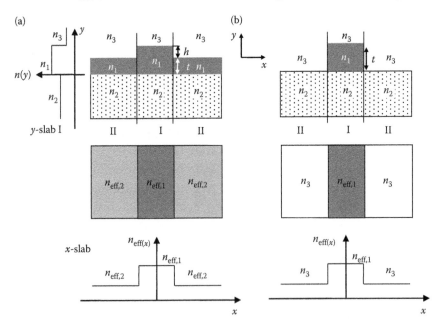

Figure 1.4 Successive approximations of a 2D waveguide by 1D slab waveguides in the EIM: (a) rib waveguide and (b) strip waveguide. The index profile $n(y)$ of the y-confined slab waveguide in the core region (region I) and the effective index distribution $n_{eff}(x)$ of the x-confined slab waveguide are also shown.

is then determined and taken as an approximation to the effective index of the 2D rectangular waveguide.

The mathematical basis of the EIM method lies in the assumption that the TE and TM semi-vectorial wave equations are separable. For example, for the quasi-TE mode, we assume that the solution for the electric field E_x in Equation 1.19 has the form

$$E_x(x,y) = X(x)Y(y). \tag{1.29}$$

Substituting this solution into Equation 1.19 and dividing by $X(x)$ $Y(y)$, we get

$$\frac{1}{X}\frac{d}{dx}\left[\frac{1}{n^2}\frac{d}{dx}\left(n^2 X\right)\right] + \frac{1}{Y}\frac{d^2 Y}{dy^2} + (n^2 k^2 - \beta_{TE}^2) = 0, \tag{1.30}$$

where $n = n(x, y)$ is the index profile of the 2D waveguide. By adding and subtracting the term $n_{eff}^2(x)k^2$ to Equation 1.30, we can separate it into two 1D wave equations (Okamoto 2000):

$$\frac{d^2 Y}{dy^2} + \left[n^2(x,y) - n_{eff}^2(x)\right]k^2 Y(y) = 0, \tag{1.31}$$

$$\frac{d}{dx}\left[\frac{1}{n_{eff}^2(x)}\frac{d}{dx}\left(n_{eff}^2(x)X\right)\right] + \left[n_{eff}^2(x)k^2 - \beta_{TE}^2\right]X(x) = 0. \tag{1.32}$$

We recognize Equation 1.31 as the TE wave equation for a y-confined slab waveguide and Equation 1.32 is the TM wave equation for an x-confined slab waveguide. We first solve Equation 1.31 in each of the core and cladding regions (regions I and II) to obtain the lateral effective index distribution $n_{eff}(x)$. The TM effective index of the equivalent x-confined slab waveguide is then determined by solving Equation 1.32. Alternatively, it is more convenient to determine the TM effective index of the x-confined slab waveguide by solving the wave equation in terms of the magnetic field H_y

$$\frac{d^2 H_y}{dx^2} + \left[n_{eff}^2(x)k^2 - \beta^2\right]H_y(x) = 0. \tag{1.33}$$

In general, the error in the effective index value obtained by the EIM method arises from two approximations. The first approximation is the use of the semi-vectorial wave equations to approximate the hybrid modes of the waveguide. The second approximation comes from the fact that in order for the semi-vectorial equation (1.19) or (1.20) to be separable, the index profile of the waveguide must be decomposable in the form (Chiang 1996)

$$n^2(x,y) = n_x^2(x) + n_y^2(y).$$ (1.34)

The actual index functions $n_x(x)$ and $n_y(y)$ assumed by the EIM method depend on the waveguide structure being analyzed.

The computation of the effective index of a 2D waveguide by the EIM method reduces to the solution of two 1D slab waveguides. In fact, one of the appealing features of the EIM method is that analytical solutions exist for the TE and TM modes of a 1D slab waveguide. In Table 1.2, we summarize the field solutions and characteristic equations for the TE and TM modes of a general asymmetric slab waveguide with width d and index distribution $n(x)$ shown in Figure 1.5.

Table 1.2 Summary of the Solutions and Characteristic Equations for the TE and TM Modes in an Asymmetric Slab Waveguide

TE Modes	TM Modes
Wave Equation	
$\dfrac{d^2 E_y}{dx^2} + [n^2(x)k^2 - \beta^2]E_y = 0$	$\dfrac{d^2 H_x}{dx^2} + [n^2(x)k^2 - \beta^2]H_x = 0$
Field Solution	
$E_y(x,z) = E_0 \psi(x) e^{-j\beta z}$	$H_x(x,z) = H_0 \psi(x) e^{-j\beta z}$

$$\psi(x) = \begin{cases} \cos(k_x d/2 + \theta)e^{-\gamma(x-d/2)}, & x > d/2 \\ \cos(k_x x + \theta), & -d/2 \le x \le d/2 \\ \cos(k_x d/2 - \theta)e^{\alpha(x+d/2)}, & x < -d/2 \end{cases}$$

Characteristic Equation

$$2k_x d - \varphi_1 - \varphi_2 = 2m\pi \quad (m = 0,1,2,3,\ldots)$$

$$4\theta = \varphi_1 - \varphi_2$$

$\varphi_1 = 2\tan^{-1}(\alpha/k_x)$	$\varphi_1 = 2\tan^{-1}(n_1^2\alpha/n_2^2 k_x)$
$\varphi_2 = 2\tan^{-1}(\gamma/k_x)$	$\varphi_2 = 2\tan^{-1}(n_1^2\gamma/n_3^2 k_x)$

$$\beta^2 + k_x^2 = n_1^2 k^2, \quad \beta^2 - \alpha^2 = n_2^2 k^2, \quad \beta^2 - \gamma^2 = n_3^2 k^2$$

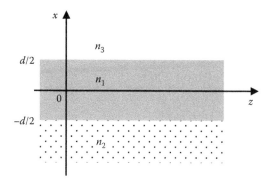

Figure 1.5   Schematic of a 1D asymmetric slab waveguide of width d, core index n_1, lower cladding index n_2, and upper cladding index n_3.

1.1.3   Waveguide dispersion

In general, the effective index of a dielectric waveguide depends on the wavelength so that light at different frequencies propagates at different velocities. This gives rise to dispersion effects such as temporal broadening of a pulse propagating in the waveguide. In a single-mode waveguide, the two main sources of this wavelength dependence (also called intramodal dispersion) are material dispersion and waveguide dispersion. Material dispersion refers to the dependence of the refractive indices of the core and cladding materials on the wavelength. Material dispersion has its physical origin in the dependence of the optical absorption of a material on frequency, so that its permittivity also depends on the frequency through the Kramers–Kronig relation. The dependence of the refractive index of a material on the wavelength can be modeled by the Sellmeier equation

$$n^2(\lambda) = 1 + \sum_i \frac{A_i}{1-(\lambda_i/\lambda)^2}. \tag{1.35}$$

The coefficients A_i and λ_i for silica and crystalline silicon are given in Table 1.3.

The second source of intramodal dispersion is waveguide dispersion, which is a structural effect and arises from the confinement of light in the waveguide core. In general, strongly confined waveguides exhibit higher waveguide dispersion due to stronger interaction of the mode with the core boundaries. In addition, in a multi-mode waveguide, there exists a third source of dispersion,

Table 1.3   Coefficients of the Sellmeier Equation for SiO_2 and Si

	SiO_2 $\lambda = 0.21\text{--}3.71\ \mu m$ at 295 K (Malitson 1965)	Si $\lambda = 1.1\text{--}5.6\ \mu m$ at 295 K (Frey et al. 2006)
A_1	0.6961663	10.67087
A_2	0.4079426	−37.10820
A_3	0.8974794	
$\lambda_1\ (\mu m)$	0.0684043	0.3045744
$\lambda_2\ (\mu m)$	0.1162414	611.2222
$\lambda_3\ (\mu m)$	9.896161	

called intermodal dispersion, which arises from the fact that different waveguide modes have different effective indices and thus propagate at different phase velocities.

The parameter used to quantify the total dependence of the effective index on wavelength is the group index, which is defined as

$$n_g = \frac{d\beta}{dk} = n_{eff} - \lambda_0 \frac{dn_{eff}}{d\lambda}. \tag{1.36}$$

From the group index, we can calculate the group velocity, which is the velocity at which a pulse with a frequency spectrum centered around λ_0 travels in the waveguide,

$$v_g = \frac{d\omega}{d\beta} = \frac{c}{n_g}. \tag{1.37}$$

The group delay experienced by a pulse after propagating a unit distance in the waveguide is given by $\tau_g = 1/v_g = d\beta/d\omega$. To express the fact that the group delay is wavelength dependent, we write τ_g in terms of a Taylor series expansion around the center wavelength λ_0,

$$\tau_g(\lambda) = \tau_g(\lambda_0) + \Delta\lambda \frac{d\tau_g}{d\lambda} + \frac{(\Delta\lambda)^2}{2} \frac{d^2\tau_g}{d\lambda^2} + \cdots \tag{1.38}$$

Defining the total chromatic dispersion of the waveguide as

$$D = \frac{d\tau_g}{d\lambda} = \frac{d}{d\lambda}\left(\frac{n_g}{c}\right) = -\frac{\lambda_0}{c}\frac{d^2 n_{eff}}{d\lambda^2}, \tag{1.39}$$

we can approximate Equation 1.38 by

$$\tau_g(\lambda) \approx \tau_g(\lambda_0) + D\Delta\lambda. \tag{1.40}$$

For a pulse of spectral width $\Delta\lambda$, we obtain from Equation 1.40 the spread in the group delay due to dispersion in the waveguide,

$$\Delta\tau = \tau_g(\lambda) - \tau_g(\lambda_0) \approx D\Delta\lambda. \tag{1.41}$$

Thus the chromatic dispersion D, typically quoted in units of ps/nm/km, gives the delay spread per unit bandwidth per unit length of the waveguide.

Figure 1.6 shows the plots of n_{eff} versus λ for the TE and TM modes of an SOI waveguide consisting of a silicon core of 250 nm thickness and 400 nm width embedded in SiO$_2$. The group index is calculated to be $n_g = 4.433$ for the TE mode and $n_g = 4.349$ for the TM mode. The chromatic dispersion of the waveguide is $D = -13.29$ ns/nm/km for the TE mode and $D = 0.972$ ns/nm/km for the TM mode. These values represent the total effects of material dispersion in the core and cladding materials as well as the structural

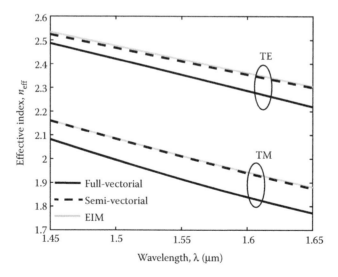

Figure 1.6 Wavelength dependence of the effective index of an SOI waveguide consisting of a silicon core with cross-sectional dimensions 400×250 nm^2 embedded in a SiO$_2$ cladding.

dispersion of the waveguide. We can also define the chromatic dispersion D_{mat} due to the material alone,

$$D_{mat} = -\frac{\lambda_0}{c}\frac{d^2 n}{d\lambda^2}, \tag{1.42}$$

where $n(\lambda)$ is the bulk index of the material. Near the 1.55 μm wavelength, the material dispersion is −0.862 ns/nm/km for Si and 21.9 ps/nm/km for SiO$_2$. Comparing these values to the total chromatic dispersion D of the waveguide reveals that waveguide dispersion plays a dominant role in silicon waveguides and indeed in high-index contrast waveguides in general.

It is evident from Figure 1.6 that the effective index also depends on the polarization. In general, the polarization dependence of the effective index arises from the birefringence of the material as well as the geometry of the waveguide. Isotropic materials such as silicon and silica do not have material birefringence, although all materials exhibit some birefringence under thermal or mechanical stress. Thus in a silicon waveguide, the dependence of the effective index on the polarization is a purely structural effect. The total birefringence of a waveguide is defined as the difference between the effective indices of two orthogonal polarization states, commonly chosen to coincide with those of the TE and TM modes:

$$B(\omega) = n_{eff}^{TM}(\omega) - n_{eff}^{TE}(\omega). \tag{1.43}$$

The frequency dependence of the birefringence gives rise to polarization mode dispersion (PMD), which is defined as

$$PMD = \frac{1}{c}\left| B(\omega) - \omega\frac{dB}{d\omega} \right| \quad (ps/km). \tag{1.44}$$

The PMD gives the differential time delay between the TE and TM components of a pulse per unit propagating distance.

1.1.4 Propagation loss

There are three main sources of loss in an optical waveguide: optical absorption in the core and cladding materials, electromagnetic scattering, and radiation leakage. Optical absorption arises from various electronic processes in the material such as atomic and molecular vibrations in glass and polymers, and interband transitions and free carrier absorption (also known as intraband transitions) in semi-conductors. Even when the waveguide is operated at a wavelength far from an optical transition, there is still some residual absorption. For example, for bulk crystalline Si, which has a bandgap of 1.1 μm, accurate measurement of the optical absorption constant in a sample with a low impurity concentration of 2×10^{12} cm^3 gives a value of 0.001 dB/cm at 1.55 μm wavelength (Steinlechner et al. 2013).

Loss due to electromagnetic scattering in an optical waveguide is caused by two mechanisms: volume or Rayleigh scattering, and surface roughness scattering. Rayleigh scattering refers to the scattering of light by small fluctuations in the refractive index caused by voids, defects, and contaminants in the material. Rayleigh scattering decreases with wavelength as λ^{-4}, and is typically much smaller than surface roughness scattering. The latter type of scattering refers to the scattering of light due the roughness of waveguide surfaces caused by fabrication processes such as deposition and etching. While the surface roughness due to deposition can be controlled to less than 1 nm, the roughness of the waveguide sidewalls due to dry etching can be as large as a few nanometers. We thus expect to have much larger scattering loss at the waveguide sidewalls than at the top and bottom surfaces of the waveguide core.

In addition to the degree of roughness, surface scattering also depends on the index contrast and how strongly light is confined in the waveguide. Several methods have been developed for estimating waveguide loss due to surface roughness scattering, ranging from the simple model of Tien based on specular reflection (Tien 1971), to more sophisticated models based on the Coupled Mode Theory (CMT) (Marcuse 1969) which take into account the statistical distribution of the roughness. However, since it is difficult to measure the roughness profiles on the sidewalls of a waveguide, the usefulness of these analyses is limited to providing broad estimates of the contributions of surface roughness scattering to the total waveguide loss.

The third major source of waveguide loss is radiation leakage. Radiation leakage arises in waveguide geometries which do not have true eigenmode solutions. The two common types of radiation loss in optical waveguides are bending loss and substrate leakage. Bending loss occurs in curved waveguides and will be discussed in more detail in Section 1.2. Substrate leakage refers to the leakage of light from a waveguide into a high-index substrate. In theory, the evanescent field of a waveguide extends indefinitely into the undercladding. The presence of a high-index substrate (such as silicon) causes evanescent coupling of light into the substrate which radiates away as loss. Substrate leakage can be minimized by increasing the thickness of the undercladding layer to provide sufficient isolation of the waveguide core from the high-index substrate.

1.2   Optical Modes in Bent Dielectric Waveguides

In a bent dielectric waveguide, the optical mode is pushed toward the outer edge of the waveguide as light propagates around the bend. As the radius of curvature increases, a portion of the evanescent tail of the mode begins to leak out in the form of radiation, resulting in bending loss. A number of techniques have been used to analyze the modes of curved optical waveguides and the associated bending loss. These techniques range from approximate analytical methods such as Marcatili's method (Marcatili 1969b), the conformal mapping method (Heiblum and Harris 1975), to rigorous numerical solutions of the wave equation in cylindrical coordinates (Rivera 1995, Lui et al. 1998, Kakihara et al. 2006). In general, approximate analytical methods give adequately accurate results for curved waveguides with low-index contrasts and large bending radii. For high-index contrast and tightly bent waveguides, the modes become highly hybridized and a full-vectorial numerical solution is required to obtain an accurate analysis of the modal characteristics.

We begin in Section 1.2.1 with an approximate analysis of bent waveguides by the conformal mapping method. Although strictly valid only for 2D structures, the conformal mapping method provides an intuitive understanding of the propagation characteristics and the mechanisms causing radiation loss in curved waveguides. For microdisk and microring structures, analytical solutions for the discrete resonant modes can be obtained by solving the 2D semi-vectorial wave equation in polar coordinates. This is the subject of Section 1.2.2. Finally, Section 1.2.3 will give a full-vectorial formulation of the problem in the three-dimensional (3D) cylindrical coordinate system (CCS) which is suitable for rigorous numerical simulations of bent waveguides.

1.2.1   Conformal transformation of bent waveguides

The idea of the conformal mapping method is to apply a coordinate transformation to the wave equation which will convert the curved boundaries of the structure in the original (x, y) coordinates into straight boundaries in the new (u, v) coordinates. Conformal mapping is based on the Cauchy–Riemann equations, which are valid for domains in a 2D plane. To apply the method to a 3D curved waveguide, we first reduce the waveguide to an equivalent 2D structure in the x–y plane using the EIM, as shown in Figure 1.7. The semi-vectorial equation governing wave propagation in the 2D bent waveguide is then given by

$$\frac{\partial^2 F}{\partial x^2} + \frac{\partial^2 F}{\partial y^2} + n^2(x,y)k^2 F = 0,$$
(1.45)

where $n(x, y)$ is the effective index distribution and

$$F(x,y) = \begin{cases} H_z, & \text{quasi-TE mode,} \\ E_z, & \text{quasi-TM mode.} \end{cases}$$
(1.46)

The conformal transformation which converts circular boundaries in the x–y plane into straight boundaries in the u–v plane is (Heiblum and Harris 1975)

$$u = \frac{R_{\text{ref}}}{2} \ln\left(\frac{x^2 + y^2}{R_{\text{ref}}^2}\right) = R_{\text{ref}} \ln\left(\frac{r}{R_{\text{ref}}}\right),$$
(1.47)

$$v = R_{\text{ref}} \tan^{-1}\left(\frac{y}{x}\right) = R_{\text{ref}}\theta,$$
(1.48)

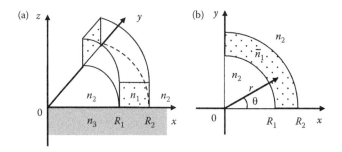

Figure 1.7 Reduction of a 3D bent waveguide in (a) to an equivalent 2D structure in (b) by the EIM. The index n_1 of the core region is replaced by the effective index $\bar{n}_1$ in (b).

where R_{ref} is a reference radius and (r, θ) are the polar coordinates of a point in the x–y plane. Applying the above transformation to the wave equation in Equation 1.45 gives

$$\frac{\partial^2 F}{\partial u^2} + \frac{\partial^2 F}{\partial v^2} + n^2(u)k^2 F = 0, \tag{1.49}$$

where the index profile of the structure in the u–v plane is given by

$$n(u) = n(x, y)e^{u/R_{\mathrm{ref}}}. \tag{1.50}$$

Figure 1.8 shows the mapping of a microdisk and a microring into straight-edge structures in the u–v plane. For the microdisk, the reference radius R_{ref} is chosen to be the microdisk radius R whereas for the microring, it is taken to be the outer radius R_2. For simplicity we have relabeled the effective index of the core as n_1 (instead of $\bar{n}_1$). Note that for each structure, the step index profile in the x–y plane is transformed into an index profile in the u–v plane that is independent of v but varying in u with exponential dependence $e^{u/R_{\mathrm{ref}}}$.

The nonlinear index profiles in Figure 1.8b and d of the transformed waveguides reveal several important aspects of wave propagation in microdisks and microrings. First, since the outer

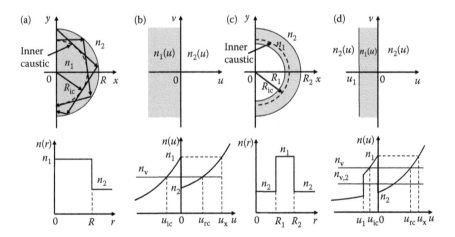

Figure 1.8 Conformal mapping of a microdisk (a) and a microring (c) in the x–y plane to straight-edge waveguides in the u–v plane, (b) and (d). The index profiles $n(r)$ in the x–y plane and $n(u)$ in the u–v plane are also shown.

cladding index exceeds the core index at the point $u = u_x$, the modes in a microdisk or microring are inherently leaky. A mode with effective index $n_v = \beta_v/k$, as indicated in Figure 1.8b and d, is confined within a "core" region $u_{ic} < u < 0$ for which $n_v < n_1(u)$. At the point $u = 0$ the wave is confined by total internal reflection (or more precisely, by frustrated total internal reflection since there is leakage into the high-index region beyond u_x). The point u_{ic}, which corresponds to the circle of radius $R_{ic} = (n_v/n_1)R_{ref}$ in the x–y plane, defines the inner caustic of the microdisk or microring (shown in Figure 1.8a and c). As illustrated in Figure 1.8a, light rays bouncing around the microdisk are tangential to the inner caustic. These modes are called whispering gallery modes. For a microring resonator, a mode behaves like a whispering gallery mode if $R_{ic} > R_1$; otherwise, it is a regular waveguide mode bounded by total internal reflection at both inner and outer walls. For example, in Figure 1.8d, the mode with effective index n_v behaves like a whispering gallery mode while the mode with effective index $n_{v,2}$ is a regular waveguide mode. Note that since the index distribution $n_1(u)$ in the core region increases toward the outer waveguide wall ($u = 0$), we expect the field distributions of the modes in a microdisk or microring to be skewed toward the outer wall, with the skewness becoming more pronounced for smaller bending radii.

At the point u_{rc} in the outer cladding region ($u > 0$) where n_v becomes smaller than $n_2(u)$, the mode becomes radiative. The point u_{rc} defines the radiation caustic and corresponds to the radius $R_{rc} = (n_v/n_2)R_{ref}$ at which the wave reaches the speed of light in the cladding medium and radiates away. The radiation caustic can also be determined by considering the microdisk in the x–y plane. As the wave travels around the microdisk, the tangential speed (or the speed in the θ direction) of the wave front must increase radially in order to maintain a straight wave front. Using the relation $n_v = n_{eff}$ $\exp(u/R_{ref}) = n_{eff}(r/R_{ref})$, where n_{eff} is the effective index in the x–y plane, we obtain for the tangential speed

$$v(r) = \frac{c}{n_{eff}} = \frac{c}{n_v}\left(\frac{r}{R_{ref}}\right). \tag{1.51}$$

The above equation shows that the tangential speed becomes equal to the speed of light in the cladding medium, c/n_2, at the radiation caustic $R_{rc} = (n_v/n_2)R_{ref}$.

To obtain the modal field distribution and propagation constant of the microring or microdisk waveguide, we can solve the wave equation (1.49) in the u–v plane. The field solution has the form

$$F(u,v) = \psi(u)e^{-j\gamma_v v},$$
(1.52)

where $\psi(u)$ is the radial field distribution and the propagation constant $\gamma_v = \beta_v - j\alpha_v$ is complex due to the leaky nature of the mode. In particular, the attenuation constant α_v gives the bending loss of the curved waveguide. Substituting (1.52) into Equation 1.49, we obtain the eigenvalue problem

$$\frac{d^2\psi}{du^2} + n^2(u)k^2\psi = \gamma_v^2\psi,$$
(1.53)

which can be solved for the propagation constant γ_v and the field distribution $\psi(u)$. Since the structure has a graded index profile $n(u)$, Equation 1.53 does not have analytical solutions and must be solved using a numerical technique such as the finite difference method. Approximate analytical solutions have also been obtained using the WKB (Wentzel–Kramers–Brillouin) method (Berglund and Gopinath 2000) or by linearizing the index profile to obtain an Airy-type equation (Chin and Ho 1998). The solution obtained in the u–v plane is then converted back to the x–y (or r–θ) plane to give

$$F(r,\theta) = \psi(R_{ref}\ln(r/R_{ref}))e^{-j\gamma_v R_{ref}\theta}.$$
(1.54)

It is evident from the above solution that γ_v gives the propagation constant along the arc length of the curved waveguide at the radial distance R_{ref} from the center. Since the effective index in the x–y plane is given by $n_{eff} = n_v(R_{eff}/r)$, we obtain the expression for the propagation constant in the x–y plane as

$$\gamma = n_{eff}k = \frac{R_{ref}}{r}\gamma_v = \frac{R_{ref}}{r}(\beta_v - j\alpha_v),$$
(1.55)

which shows that γ, and hence n_{eff}, vary as $1/r$. It is typical to define the effective propagation constant of a microdisk or microring as the value of γ at the effective radius R_{eff},

$$\gamma_{eff} = \frac{R_{ref}}{R_{eff}}(\beta_v - j\alpha_v),$$
(1.56)

where R_{eff} is the radial distance to the centroid of the intensity distribution of the mode (Rowland and Love 1993):

$$R_{eff} = \frac{\int_0^{R_{rc}} |\psi(r)|^2 \, dr}{\int_0^{R_{rc}} \frac{1}{r} |\psi(r)|^2 \, dr}. \tag{1.57}$$

For microrings, the effective radius is very close to the average radius, $R_{avg} = (R_1 + R_2)/2$, and in most cases can be approximated by this value.

1.2.2 Resonant modes in microdisks and microrings

In the previous section we have assumed that the waveguide is unbounded in the propagation direction (i.e., the v direction in the u–v plane or the θ direction in the r–θ plane). As a result, the propagation constant β_v is a continuous function of frequency. In a microdisk or microring resonator, the imposition of the periodic boundary condition in θ (or v) gives rise to discrete eigenmodes at discrete frequencies, which are the resonant modes of the structure. For 2D microdisks and microrings, we can obtain analytical solutions for the field distributions of the eigenmodes and their resonant frequencies by solving the semi-vectorial wave equation in polar coordinates subject to the periodic boundary condition.

We consider first the solution for a microdisk with radius R. As in the previous section, we assume that the 3D microdisk structure has been reduced to an equivalent 2D structure using the EIM, with the effective index distribution in the r–θ plane given by

$$n(r) = \begin{cases} n_1, & r \le R, \\ n_2, & r > R. \end{cases} \tag{1.58}$$

In polar coordinates, the equation governing wave propagation in the microdisk is

$$\nabla^2 F + n^2 k^2 F = \frac{\partial^2 F}{\partial r^2} + \frac{1}{r} \frac{\partial F}{\partial r} + \frac{1}{r^2} \frac{\partial^2 F}{\partial \theta^2} + n^2(r) k^2 F = 0, \tag{1.59}$$

where $k = \omega/c$ and

$$F(r,\theta) = \begin{cases} H_z, & \text{quasi-TE mode,} \\ E_z, & \text{quasi-TM mode.} \end{cases}$$

Equation 1.59 can be solved using the method of separation of variables. Letting $F(r,\theta) = \psi(r)\Theta(\theta)$, we separate the equation to get

$$\frac{d^2\Theta}{d\theta^2} + \beta_\theta^2 \Theta = 0, \tag{1.60}$$

$$\frac{d^2\psi}{dr^2} + \frac{1}{r}\frac{d\psi}{dr} + \left[n^2(r)k^2 - \frac{\beta_\theta^2}{r^2} \right]\psi = 0, \tag{1.61}$$

where β_θ^2 is the constant of separation. The solution of Equation 1.60 subject to the periodic boundary condition $\Theta(\theta) = \Theta(\theta + 2\pi)$ is

$$\Theta_m(\theta) = e^{\pm jm\theta}, \quad m = 0, 1, 2, 3, \ldots \tag{1.62}$$

with the separation constant given by $\beta_\theta^2 = m^2$. The integer m denotes the azimuthal number of the resonant mode. Substituting $\beta_\theta = m$ into Equation 1.61, we get

$$\frac{d^2\psi}{dr^2} + \frac{1}{r}\frac{d\psi}{dr} + \left[n^2(r)k^2 - \frac{m^2}{r^2} \right]\psi = 0, \tag{1.63}$$

which is the Bessel equation of order m. Assuming waves in the cladding region propagate outward in the radial direction, we write the solution to Equation 1.63 as

$$\psi(r) = \begin{cases} C_1 J_m(n_1 kr), & r \leq R, \\ C_2 H_m^{(2)}(n_2 kr), & r > R, \end{cases} \tag{1.64}$$

where J_m and $H_m^{(2)}$ are the Bessel function and Hankel function of the second kind, respectively, of order m. Since ψ must be continuous at the dielectric interface at $r = R$, we obtain the relation for the amplitude coefficients,

$$C_2 = \frac{J_m(n_1 kR)}{H_m^{(2)}(n_2 kR)} C_1. \tag{1.65}$$

In addition, we also require that $E_\theta \propto (1/n^2)d\psi/dr$ be continuous at $r = R$ for the TE mode, and $H_\theta \propto d\psi/dr$ be continuous at $r = R$ for the TM mode. Enforcing the above field continuity conditions and making use of Equation 1.65, we obtain the characteristic equation for the microdisk[*]

$$\frac{J_m'(n_1 kR)}{J_m(n_1 kR)} = \frac{s H_m^{(2)'}(n_2 kR)}{H_m^{(2)}(n_2 kR)}, \tag{1.66}$$

where $s = n_1/n_2$ for TE and $s = n_2/n_1$ for TM. Solution of the above equation gives the eigenvalues k_{mn}, where $n = 1, 2, 3, \ldots$ represents the radial mode number. Since the modes in a microdisk are leaky, k_{mn} is a complex number. The characteristic frequency of mode (m, n) is thus also complex, which can be expressed as

$$\omega_{mn} = \omega_{mn}' + j\omega_{mn}'' = ck_{mn}. \tag{1.67}$$

Using the above expression, we can write the time-dependent solution for the resonant mode (m, n) in the microdisk as

$$F(r,\theta,t) = \psi_{mn}(r) e^{\pm jm\theta} e^{j\omega_{mn}t} = \psi_{mn}(r) e^{j(\omega_{mn}'t \pm m\theta)} e^{-\omega_{mn}''t}, \tag{1.68}$$

where the plus and minus signs correspond to the counterclockwise and clockwise propagating modes, respectively. In Equation 1.68, ω_{mn}' is the resonant frequency of mode (m, n) and ω_{mn}'' gives the temporal rate of field decay in the cavity due to bending loss. The angular velocity of the phase front of mode (m, n) is $v_{\theta,mn} = \omega_{mn}'/m$. We may define the effective index of mode (m, n) from the tangential velocity $(rv_{\theta,mn})$ of the wave at the effective radial distance R_{eff} from the microdisk center,

$$n_{eff,mn} = \frac{c}{R_{eff}v_{\theta,mn}} = \frac{m\lambda_{mn}}{2\pi R_{eff}}, \tag{1.69}$$

where $\lambda_{mn} = 2\pi c/\omega_{mn}'$ is the resonant wavelength of mode (m, n). The above equation can be recast in the form

$$\frac{m\lambda_{mn}}{n_{eff,mn}} = 2\pi R_{eff}, \tag{1.70}$$

---

[*] At nonresonant frequencies, the azimuthal equation (1.60) does not have to satisfy the periodic boundary condition so it must be treated as an open boundary problem. In this case the separation constant $\beta_\theta = v$ is an unknown complex number and we must replace J_m and $H_m^{(2)}$ by J_v and $H_v^{(2)}$, respectively, in the general solution for ψ in (1.64). The characteristic equation (1.66) is then an equation in terms of the complex order v of the Bessel and Hankel functions. Solution of the characteristic equation involving Bessel and Hankel functions of complex orders has been investigated for bent waveguides in Hiremath et al. (2005).

which indicates that at resonance, the effective microdisk circumference $(2\pi R_{\text{eff}})$ must be equal to an integer multiple of the guided wavelength.

The attenuation constant of the mode due to bending loss can be obtained by converting the temporal decay factor in Equation 1.68 to an angular attenuation factor:

$$\exp(-\omega''_{mn}t) = \exp\left(-\omega''_{mn}\theta/v_{\theta,mn}\right) = \exp\left(-m\omega''_{mn}\theta/\omega'_{mn}\right). \qquad (1.71)$$

From the above expression, we obtain the power attenuation constant α_{mn} along the arc length at a radial distance R_{eff} from the microdisk center to be

$$\alpha_{mn} = \frac{2m\omega''_{mn}}{\omega'_{mn}R_{\text{eff}}}. \qquad (1.72)$$

Another parameter of interest is the quality factor (Q factor) of the microdisk resonator, which is defined as 2π times the ratio of the time-averaged energy stored to the energy loss per cycle (Jackson 1999):

$$Q = 2\pi \frac{\text{Average stored energy}}{\text{Power loss} \times \text{Optical period}} = \omega'_{mn} \frac{\text{Average stored energy}}{\text{Power loss}}. \qquad (1.73)$$

Since the instantaneous energy density in the microdisk is proportional to the square of the real field in Equation 1.68, we can express the total stored energy in the resonator at time t as

$$U(t) = U_0 \cos^2(\omega'_{mn}t)e^{-2\omega''_{mn}t}, \qquad (1.74)$$

where U_0 is the initial stored energy. Assuming that the rate of energy decay is much smaller than the angular frequency of the optical field $(2\omega''_{mn} \ll \omega'_{mn})$, the stored energy in the microdisk averaged over one optical cycle is

$$\langle U \rangle = \frac{1}{2}U_0 e^{-2\omega''_{mn}t}. \qquad (1.75)$$

Since the power loss is just the rate of energy decay,

$$P_L = -\frac{d\langle U \rangle}{dt} = 2\omega''_{mn}\langle U \rangle, \qquad (1.76)$$

we obtain the quality factor associated with the resonant mode (m, n) as

$$Q = \omega'_{mn} \frac{\langle U \rangle}{P_L} = \frac{\omega'_{mn}}{2\omega''_{mn}}. \tag{1.77}$$

If we define the cavity lifetime (or photon lifetime) τ_{mn} as the time it takes for the stored energy to decay to $1/e$ of its initial value, we have from Equations 1.75 and 1.77 that $\tau_{mn} = 1/2\omega''_{mn} = Q/\omega'_{mn}$. Thus the Q factor (divided by 2π) gives the cavity lifetime in terms of the number of optical cycles.

Solutions of the resonant modes in a microring resonator can be obtained in a similar manner as for the microdisk. The only difference lies in the solution of Equation 1.63 for the radial field distribution. For a 2D microring with inner radius R_1, outer radius R_2, and effective index distribution

$$n(r) = \begin{cases} n_1, & R_1 \leq r \leq R_2, \\ n_2, & r < R_1 \text{ and } r > R_2, \end{cases} \tag{1.78}$$

the solution for the radial field distribution has the form

$$\psi(r) = \begin{cases} C_1 I_m(n_2 k r), & r < R_1, \\ C_2 J_m(n_1 k r) + C_3 Y_m(n_1 k r), & R_1 \leq r \leq R_2, \\ C_4 H_m^{(2)}(n_2 k r), & r > R_2, \end{cases} \tag{1.79}$$

where Y_m and I_m are, respectively, the Neumann function and the modified Bessel function of the first kind of order m. Again the Hankel function is chosen as the solution in the outer cladding region, $r > R_2$, to allow for radiation leakage. By matching the tangential fields at the inner and outer radii, we obtain the following characteristic equation for the microring:

$$\frac{s I'_m(n_2 k R_1) J_m(n_1 k R_1) - I_m(n_2 k R_1) J'_m(n_1 k R_1)}{s H_m^{(2)'}(n_2 k R_2) J_m(n_1 k R_2) - H_m^{(2)}(n_2 k R_2) J'_m(n_1 k R_2)}$$
$$= \frac{s I'_m(n_2 k R_1) Y_m(n_1 k R_1) - I_m(n_2 k R_1) Y'_m(n_1 k R_1)}{s H_m^{(2)'}(n_2 k R_2) Y_m(n_1 k R_2) - H_m^{(2)}(n_2 k R_2) Y'_m(n_1 k R_2)}, \tag{1.80}$$

where $s = n_1/n_2$ for TE modes and $s = n_2/n_1$ for TM modes.

Numerical results for the resonant modes in silicon microrings and microdisks obtained using the above analysis are shown in

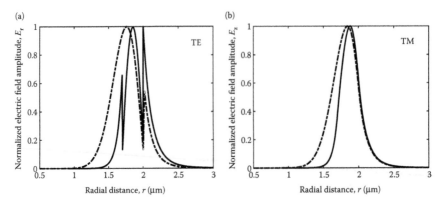

Figure 1.9 Transverse electric field amplitude of the resonant mode $(m, n) = (23, 1)$ in a silicon microdisk with 2 μm radius (dashed line) and a silicon microring with $R_1 = 1.7$ μm and $R_2 = 2$ μm (solid line): (a) E_r field of the TE mode, (b) E_z field of the TM mode.

Figures 1.9 and 1.10. In all the plots we assume a Si waveguide core of thickness of 250 nm embedded in a SiO_2 cladding. Figure 1.9a and b show the transverse electric field distributions of the TE and TM resonant modes with $m = 23$, $n = 1$, in a microdisk with a 2 μm radius and a microring with $R_1 = 1.7$ μm and $R_2 = 2$ μm. A slight skewing of the modes toward the outer edge is noticeable, especially for the TM mode. Figure 1.10a and b show the effective indices of the TE and TM resonant modes as functions of the resonant wavelength for microrings with a fixed average radius of 2 μm and microring width varying from 250 nm to 1 μm. In both plots, lines of constant azimuthal mode number m and constant ring width are also shown. We observe that the effective index of the microring exhibits strong dispersion, especially for small ring widths. As the ring width increases, the microring behaves more like a whispering gallery mode resonator so its dispersion characteristic approaches that of a microdisk, whose effective index is shown by the bold dark line in both plots. Figure 10c and d show the dependence of the Q factor of the microring on the resonant wavelength and the ring width. The bold dark line in each plot indicates the Q factor of the microdisk, which serves as the limiting case for microrings with large ring widths. From these plots we observe that for a fixed ring width, the Q factor decreases with wavelength since the mode becomes less confined, leading to higher radiation leakage due to bending.

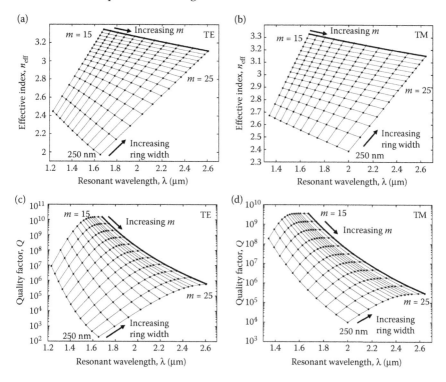

Figure 1.10 Effective indices of the resonant modes of a silicon microring with core thickness of 250 nm, average radius of 2 μm, and ring width varying from 250 nm to 1 μm in steps of 50 nm: (a) TE mode, (b) TM mode. Quality factors of (c) TE and (d) TM resonant modes in the microrings. The bold dark lines in all the plots are the results for a silicon microdisk with 2.5 μm radius.

Figure 1.11 shows the dependence of the bending loss of a silicon microring on the average ring radius for TE and TM resonant modes near the 1.55 μm wavelength. The microring width is fixed at 400 nm. The bending loss is seen to exhibit an exponential dependence on the bending radius. This exponential dependence has also been predicted by Marcuse's approximate analytical formula for loss in a bent slab waveguide (Marcuse 1972). Due to the high-index contrast of the SOI material system, the bending loss of silicon microrings is seen to remain low for very tight bends with radii down to 2.5 μm. For comparison we also show the bending loss of SiN microrings with SiO$_2$ cladding. The SiN core thickness is assumed to be 400 nm and the ring width is 800 nm. Since the SiN/SiO$_2$ material system has a lower index contrast, the bending

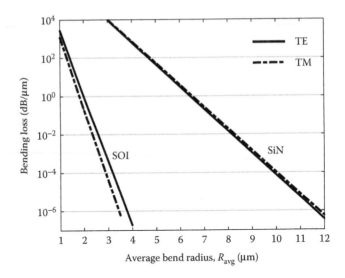

Figure 1.11   Bending loss as a function of the average bend radius of SOI and SiN microrings. The dimensions of the silicon ring waveguide are 250 × 400 nm (height × width), while those of the SiN ring waveguide are 400 × 800 nm. The bending losses are computed for resonant modes near the 1.55 μm wavelength.

loss lines for SiN microrings are shifted to larger radii compared to those of SOI microrings, although compact bends with radii less than 10 μm can still be achieved with low loss in SiN.

1.2.3   Full-vectorial analysis of bent waveguides

The approximate analytical solutions in the previous sections are useful for gaining an intuitive understanding of the characteristics and behavior of wave propagation in bent waveguides. For an accurate determination of the modes and bending loss in a 3D structure, however, we must resort to a full-vectorial analysis due to the high degree of hybridization of the modes. This effect is especially pronounced in tight bends in high-index contrast materials such as SOI, where low loss waveguide bends and microrings with radii approaching 1 μm have been demonstrated (Vlasov and McNab 2004, Xu et al. 2008, Prabhu et al. 2010). A full vectorial analysis of curved waveguides requires a numerical solution of the vector wave equation, which is typically formulated in a local cylindrical coordinate system (CCS) centered about the waveguide core.

We consider a bent waveguide with inner radius R_1, outer radius R_2, and cross-sectional index distribution $n(r, z)$ as shown

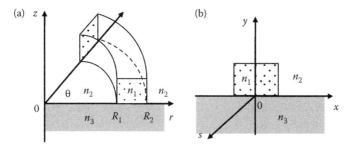

Figure 1.12   Waveguide bend in (a) the global cylindrical coordinate system (r, θ, z) and (b) the local cylindrical coordinate system (x, y, s).

in Figure 1.12a. In the CCS, we decompose the electric field in the waveguide into a transverse and angular component as follows:

$$\mathcal{E}(r,\theta,z) = \mathbf{E}_t(r,\theta,z) + E_\theta(r,\theta,z)\hat{\theta}, \tag{1.81}$$

where $\mathbf{E}_t = E_r\hat{\mathbf{r}} + E_z\hat{\mathbf{z}}$ and $\hat{\theta}$ is the unit vector in the angular direction. Substituting the above expression into the wave equation in Equation 1.9,

$$\nabla^2\mathcal{E} + n^2k^2\mathcal{E} = \nabla(\nabla\cdot\mathcal{E}), \tag{1.82}$$

we obtain the following equation for the transverse field component $\mathbf{E}_t$:[*]

$$\nabla_t^2\mathbf{E}_t + \frac{1}{r^2}\frac{\partial^2\mathbf{E}_t}{\partial\theta^2} + n^2k^2\mathbf{E}_t = \nabla_t(\nabla_t\cdot\mathbf{E}_t) - \nabla_t\left[\frac{1}{n^2}\nabla_t\cdot(n^2\mathbf{E}_t)\right], \tag{1.83}$$

where

$$\nabla_t = \frac{\partial}{\partial r}\hat{\mathbf{r}} + \frac{\partial}{\partial z}\hat{\mathbf{z}} \quad \text{and} \quad \nabla_t^2 = \frac{1}{r}\frac{\partial}{\partial r}\left(r\frac{\partial}{\partial r}\right) + \frac{\partial^2}{\partial z^2}.$$

It is more convenient to solve Equation 1.83 in a local CCS (x, y, s) centered about the waveguide core, as shown in Figure 1.12b. The transformations from the global CCS (r, θ, z) to the local CCS are given by (Cheng et al. 1990)

---

[*] Equation 1.83 is similar to Equation 1.13 except that it is expressed in cylindrical coordinates.

$$r \rightarrow x + R, \theta \rightarrow s/R, z \rightarrow y, \tag{1.84}$$

where $R = (R_1 + R_2)/2$ is the average bending radius. We also rename the transverse field components $E_r \rightarrow E_x$, $E_z \rightarrow E_y$ and assume an $e^{-j\beta s}$ dependence in the propagation direction. The transverse field can now be written as

$$\mathbf{E}_t = \left[E_x(x,y)\hat{\mathbf{x}} + E_y(x,y)\hat{\mathbf{y}} \right] e^{-j\beta s}. \tag{1.85}$$

Applying the above coordinate transformations to Equation 1.83, we obtain a system of equations for the transverse electric field components which can be put in matrix form as follows (Lui et al. 1998, Feng et al. 2002):

$$\begin{bmatrix} P_{xx} & P_{xy} \\ P_{yx} & P_{yy} \end{bmatrix} \begin{bmatrix} E_x \\ E_y \end{bmatrix} = \beta^2 \begin{bmatrix} E_x \\ E_y \end{bmatrix}, \tag{1.86}$$

where

$$P_{xx}E_x = \frac{1}{\rho^2} \frac{\partial}{\partial x} \left[\frac{\rho}{n^2} \frac{\partial}{\partial x} \left(\rho n^2 E_x \right) \right] + \frac{\partial^2 E_x}{\partial y^2} + n^2 k^2 E_x, \tag{1.87}$$

$$P_{xy}E_y = \frac{1}{\rho^2} \frac{\partial}{\partial x} \left[\frac{\rho^2}{n^2} \frac{\partial}{\partial y} \left(n^2 E_y \right) \right] - \frac{\partial^2 E_y}{\partial y \partial x}, \tag{1.88}$$

$$P_{yy}E_y = \frac{1}{\rho} \frac{\partial}{\partial x} \left(\rho \frac{\partial E_y}{\partial x} \right) + \frac{\partial}{\partial y} \left[\frac{1}{n^2} \frac{\partial}{\partial y} \left(n^2 E_y \right) \right] + n^2 k^2 E_y, \tag{1.89}$$

$$P_{yx}E_x = \frac{\partial}{\partial y} \left[\frac{1}{\rho n^2} \frac{\partial}{\partial x} \left(\rho n^2 E_x \right) \right] - \frac{1}{\rho} \frac{\partial}{\partial x} \left(\rho \frac{\partial E_x}{\partial y} \right), \tag{1.90}$$

and $\rho = 1 + x/R$. In the limit $R \rightarrow \infty$, the above equations reduce to Equations 1.15 through 1.18 for a straight waveguide.

We can also formulate a similar set of equations in terms of the transverse magnetic field. From the wave equation (1.22) for the magnetic field,

$$\nabla^2 \mathcal{H} + n^2 k^2 \mathcal{H} = -\frac{1}{n^2} \nabla n^2 \times (\nabla \times \mathcal{H}), \tag{1.91}$$

we make the substitution $\mathcal{H}(r,\theta,z) = \mathbf{H}_t(r,\theta,z) + H_\theta(r,\theta,z)\hat{\theta}$ to obtain the following equation for the transverse field component in the global CCS (Lui et al. 1998):

$$\nabla_t^2 \mathbf{H}_t + \frac{1}{r^2}\frac{\partial^2 \mathbf{H}_t}{\partial\theta^2} + n^2 k^2 \mathbf{H}_t = -\frac{1}{n^2}(\nabla_t n^2) \times (\nabla_t \times \mathbf{H}_t). \tag{1.92}$$

Writing the transverse magnetic field in the local CCS as

$$\mathbf{H}_t = \left[H_x(x,y)\hat{x} + H_y(x,y)\hat{y}\right]e^{-j\beta s}, \tag{1.93}$$

and applying the coordinate transformations in Equation 1.84 to Equation 1.92, we obtain the matrix equation (Lui et al. 1998, Xiao et al. 2009)

$$\begin{bmatrix} Q_{xx} & Q_{xy} \\ Q_{yx} & Q_{yy} \end{bmatrix} \begin{bmatrix} H_x \\ H_y \end{bmatrix} = \beta^2 \begin{bmatrix} H_x \\ H_y \end{bmatrix}, \tag{1.94}$$

where

$$Q_{xx}H_x = \frac{\partial}{\partial x}\left[\rho\frac{\partial(\rho H_x)}{\partial x}\right] + \rho^2 n^2 \frac{\partial}{\partial y}\left(\frac{1}{n^2}\frac{\partial H_x}{\partial y}\right) + \rho^2 n^2 k^2 H_x, \tag{1.95}$$

$$Q_{xy}H_y = \frac{\partial}{\partial x}\left(\rho^2\frac{\partial H_y}{\partial x}\right) - \rho^2 n^2 \frac{\partial}{\partial y}\left(\frac{1}{n^2}\frac{\partial H_y}{\partial x}\right), \tag{1.96}$$

$$Q_{yy}H_y = \rho n^2 \frac{\partial}{\partial x}\left(\frac{\rho}{n^2}\frac{\partial H_y}{\partial x}\right) + \rho^2 \frac{\partial^2 H_y}{\partial y^2} + \rho^2 n^2 k^2 H_y, \tag{1.97}$$

$$Q_{yx}H_x = \rho^2 \frac{\partial^2 H_x}{\partial x \partial y} - \rho^2 n^2 \frac{\partial}{\partial x}\left(\frac{1}{n^2}\frac{\partial H_x}{\partial y}\right), \tag{1.98}$$

and $\rho = 1 + x/R$. The above equations simplify to Equations 1.25 through 1.28 for a straight waveguide in the limit $R \to \infty$.

Equations 1.86 and 1.94 can be solved using the finite difference method (Lui et al. 1998) for the transverse field distributions and the complex propagation constant β. Alternatively, one can also solve Equation 1.82 or 1.91 directly in cylindrical coordinates using the finite element method (Kakihara et al. 2006). Due to the leaky nature of the modes, care should be taken to apply appropriate absorbing boundary conditions (ABCs) at the boundaries of the computational domain to properly absorb the radiating waves. In particular, the perfectly matched layer ABCs have been shown to be effective in both the finite difference and finite element solutions (Feng et al. 2002, Kakihara et al. 2006).

Figure 1.13a and b show the transverse magnetic field distributions in the local CCS of the lowest quasi-TM mode of an SOI bent waveguide (Prabhu et al. 2010). The waveguide dimensions are 340×300 nm^2 (height $\times$ width) and the average bending radius is 1 µm. The skewing of the field distributions toward the outer radius (in the positive x direction) is apparent. Figure 1.13c and d show the roundtrip loss and the Q factor at the 1.55 µm wavelength as functions

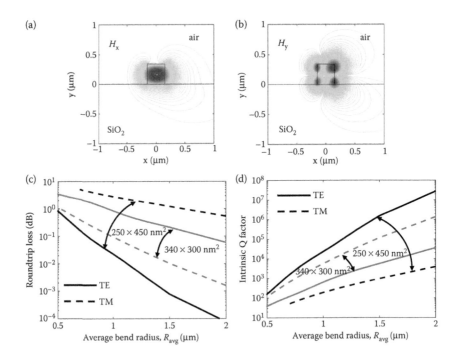

Figure 1.13 (a) and (b) Transverse magnetic field components H_x and H_y of the quasi-TM mode of an SOI curved waveguide with average radius of 1.0 µm and waveguide dimensions of 340 nm $\times$ 300 nm (height $\times$ width). Dependence of the roundtrip bending loss (c) and the intrinsic Q factor (d) of SOI curved waveguides on the bending radius at 1.55 µm wavelength for two sets of waveguide dimensions, 340×300 nm^2 and 250×450 nm^2. (From Prabhu, A.M., et al., 2010, *IEEE Photonics J.*, 2(3): 436–444.)

of the bending radius for SOI microrings with two waveguide aspect ratios (height × width): 340×300 nm^2 and 250×450 nm^2. We observe that similar to the trend shown in Figure 1.11, the bending loss decreases exponentially with increasing radius for both the TE and TM polarizations. Note also the strong dependence of the bending loss and the Q factor on the polarization. For example, for a 2-μm radius microring with aspect ratio 250×450 nm^2, the bending losses and Q factors of the TE and TM modes differ by as much as 4 orders of magnitude. Also noteworthy from these plots is the fact that fairly high intrinsic Q values can still be achieved with extremely small microring resonators. For example, for the 1-μm radius microring with aspect ratio 250×450 nm^2, the intrinsic Q factor associated with bending loss exceeds 20,000 for the TE mode.

1.3   Coupling of Waveguide Modes in Space

When two dielectric waveguides are brought into close proximity of each other, the evanescent tail of the modal field distribution of each waveguide interacts with the other waveguide core, resulting in coupling and power exchange between the two waveguides. This evanescent coupling of power is the primary means by which light is coupled into a microring resonator from an external waveguide, or between two microring resonators. Evanescent coupling between two dielectric waveguides is analyzed by means of the Coupled Mode Theory (CMT), which describes the coupling of modes of the two waveguides as they propagate in space. The CMT equations can be formulated from Maxwell's equations using a number of approaches such as perturbation theory (Yariv 1973), variational approach (Haus et al. 1987), and reciprocity (Chuang 1987a,b). In this section we will adopt the reciprocity approach to formulate the equations for the coupling of waveguide modes in space and derive the solution for a pair of evanescently coupled waveguides.

1.3.1   The coupled mode equations

We begin by deriving a general reciprocity relation for the electric and magnetic fields in two media (or structures) described by the z-invariant permittivity functions $\varepsilon_a(x, y)$ and $\varepsilon_b(x, y)$ (Chuang 1987a). The fields $\mathbf{E}_a$, $\mathbf{H}_a$ and $\mathbf{E}_b$, $\mathbf{H}_b$ in media a and b satisfy the Maxwell's equations

$$\nabla \times \mathbf{E}_a = -j\omega\mu_0\mathbf{H}_a, \tag{1.99}$$

$$\nabla \times \mathbf{H}_a = j\omega\varepsilon_a(x,y)\mathbf{E}_a, \tag{1.100}$$

$$\nabla \times \mathbf{E}_b = -j\omega\mu_0\mathbf{H}_b, \tag{1.101}$$

$$\nabla \times \mathbf{H}_b = j\omega\varepsilon_b(x,y)\mathbf{E}_b. \tag{1.102}$$

By expanding the expression $\nabla \cdot (\mathbf{E}_a \times \mathbf{H}_b - \mathbf{E}_b \times \mathbf{H}_a)$ and making use of the above equations, we get

$$\nabla \cdot (\mathbf{E}_a \times \mathbf{H}_b - \mathbf{E}_b \times \mathbf{H}_a) = -j\omega(\varepsilon_b - \varepsilon_a)\mathbf{E}_a \cdot \mathbf{E}_b. \tag{1.103}$$

Taking the integral of Equation 1.103 over the cross-sectional area in the x–y plane, we obtain with the help of the divergence theorem,

$$\frac{\partial}{\partial z}\int (\mathbf{E}_a \times \mathbf{H}_b - \mathbf{E}_b \times \mathbf{H}_a) \cdot \hat{z}dxdy = -j\omega\int (\varepsilon_b - \varepsilon_a)\mathbf{E}_a \cdot \mathbf{E}_b dxdy. \tag{1.104}$$

The above reciprocity relation holds for any two sets of fields $\mathbf{E}_a$, $\mathbf{H}_a$ and $\mathbf{E}_b$, $\mathbf{H}_b$ satisfying Maxwell's equations in any two media $\varepsilon_a(x,y)$ and $\varepsilon_b(x,y)$. Physically, it describes the interaction between the two sets of fields through the polarization current $\mathbf{J}_j = j\omega(\varepsilon_i - \varepsilon_j)\mathbf{E}_j$ induced in each medium $i, j = \{a, b\}$.

We use the reciprocity relation in Equation 1.104 to formulate the CMT equations for a system of two coupled parallel waveguides separated by a gap s, as shown in Figure 1.14c. The permittivity functions of the isolated waveguides are given by $\varepsilon_1(x,y)$ and $\varepsilon_2(x,y)$, as shown in Figure 1.14a and b, and that of the coupled waveguide pair is $\varepsilon(x,y)$. Let us choose medium a to represent the waveguide pair, $\varepsilon_a = \varepsilon(x,y)$, and medium b to be waveguide 1 in isolation, $\varepsilon_b = \varepsilon_1(x,y)$. In the coupled

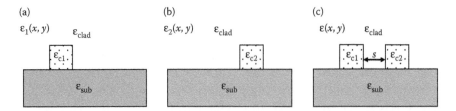

Figure 1.14 (a) and (b) Cross sections of isolated waveguides 1 and 2 described by permittivity functions $\varepsilon_1(x, y)$ and $\varepsilon_2(x, y)$. (c) Coupled waveguide pair with permittivity function $\varepsilon(x, y)$.

waveguides, we approximate the forward-propagating fields in terms of a linear superposition of the modes of waveguides 1 and 2,[*]

$$\mathbf{E}_a = a_1(z)\mathbf{e}^{(1)}(x,y) + a_2(z)\mathbf{e}^{(2)}(x,y), \tag{1.105}$$

$$\mathbf{H}_a = a_1(z)\mathbf{h}^{(1)}(x,y) + a_2(z)\mathbf{h}^{(2)}(x,y), \tag{1.106}$$

where $\mathbf{e}^{(1)}$, $\mathbf{h}^{(1)}$ and $\mathbf{e}^{(2)}$, $\mathbf{h}^{(2)}$ are the electric and magnetic field distributions of the modes of the isolated waveguides 1 and 2 with the power normalization

$$\frac{1}{2}\int \mathbf{e}^{(i)} \times \mathbf{h}^{(i)} \cdot \hat{z}\,dx\,dy = 1, \quad i = \{1,2\}.$$

In subsequent derivations, it is convenient to decompose the modal fields into a transverse and longitudinal component as

$$\mathbf{e}^{(1)}(x,y) = \mathbf{e}_t^{(1)} + \hat{z}e_z^{(1)}, \tag{1.107}$$

$$\mathbf{h}^{(1)}(x,y) = \mathbf{h}_t^{(1)} + \hat{z}h_z^{(1)}, \tag{1.108}$$

with similar expressions for $\mathbf{e}^{(2)}$, $\mathbf{h}^{(2)}$. In medium b, we take the fields $\mathbf{E}_b$, $\mathbf{H}_b$ to be the backward-propagating fields in waveguide 1:

$$\mathbf{E}_b = \mathbf{e}^{(1)}(x,y)e^{j\beta_1 z} = (\mathbf{e}_t^{(1)} - \hat{z}e_z^{(1)})e^{j\beta_1 z}, \tag{1.109}$$

---

[*] To be more accurate, we should express only the transverse components of $\mathbf{E}_a$ and $\mathbf{H}_a$ by a linear superposition of the transverse fields of the modes of waveguides 1 and 2:

$$\mathbf{E}_t^{(a)} = a_1(z)\mathbf{e}_t^{(1)}(x,y) + a_2(z)\mathbf{e}_t^{(2)}(x,y),$$

$$\mathbf{H}_t^{(a)} = a_1(z)\mathbf{h}_t^{(1)}(x,y) + a_2(z)\mathbf{h}_t^{(2)}(x,y).$$

The longitudinal field components of $\mathbf{E}_a$ and $\mathbf{H}_a$ are then derived from Maxwell's equations to give (Chuang 1987a)

$$E_z^{(a)} = a_1(z)\frac{\varepsilon_1}{\varepsilon}e_z^{(1)}(x,y) + a_2(z)\frac{\varepsilon_2}{\varepsilon}e_z^{(2)}(x,y),$$

$$H_z^{(a)} = a_1(z)h_z^{(1)}(x,y) + a_2(z)h_z^{(2)}(x,y).$$

The above field expressions lead to a slight modification to Equation 1.114 for the coupling term:

$$K_{i,j} = \frac{\omega}{4}\int\int (\varepsilon - \varepsilon_j)\left[\mathbf{e}_t^{(i)} \cdot \mathbf{e}_t^{(j)} - \frac{\varepsilon_i}{\varepsilon}e_z^{(i)}e_z^{(j)}\right]dx\,dy.$$

$$\mathbf{H}_b = \mathbf{h}^{(1)}(x,y)e^{j\beta_1 z} = (-\mathbf{h}_t^{(1)} + \hat{z}h_z^{(1)})e^{j\beta_1 z}, \tag{1.110}$$

where β_1 is the propagation constant of waveguide 1. By substituting the two sets of fields in Equations 1.105, 1.106, and 1.109, 1.110 into the reciprocity relation (1.104), we obtain after some simplification

$$\frac{da_1}{dz} + C\frac{da_2}{dz} = j(K_{11} + \beta_1)a_1(z) + j(K_{21} + C\beta_1)a_2(z), \tag{1.111}$$

where

$$C = \frac{C_{12} + C_{21}}{2}, \tag{1.112}$$

$$C_{i,j} = \frac{1}{2}\int \mathbf{e}_t^{(i)} \times \mathbf{h}_t^{(j)} \cdot \hat{z}\,dx\,dy, \qquad \{i,j\} = \{1,2\}, \tag{1.113}$$

$$K_{i,j} = \frac{\omega}{4}\int(\varepsilon - \varepsilon_j)\mathbf{e}^{(i)} \cdot \mathbf{e}^{(j)}dx\,dy$$
$$= \frac{\omega}{4}\int(\varepsilon - \varepsilon_j)\left[\mathbf{e}_t^{(i)} \cdot \mathbf{e}_t^{(j)} - e_z^{(i)}e_z^{(j)}\right]dx\,dy, \quad \{i,j\} = \{1,2\}. \tag{1.114}$$

In a similar manner, if we choose medium a to be the waveguide pair, $\varepsilon_a = \varepsilon(x, y)$, and medium b to be waveguide 2, $\varepsilon_b = \varepsilon_2(x, y)$, we obtain a second equation in terms of the mode amplitudes a_1 and a_2,

$$C\frac{da_1}{dz} + \frac{da_2}{dz} = j(K_{12} + C\beta_2)a_1(z) + j(K_{22} + \beta_2)a_2(z), \tag{1.115}$$

where β_2 is the propagation constant of waveguide 2. The coupling terms K_{12} and K_{22} are given by Equation 1.114. Equations 1.111 and 1.115 can be combined as

$$\mathbf{C}\frac{d}{dz}\begin{bmatrix} a_1(z) \\ a_2(z) \end{bmatrix} = -j(\mathbf{K} + \mathbf{BC})\begin{bmatrix} a_1(z) \\ a_2(z) \end{bmatrix}, \tag{1.116}$$

where

$$\mathbf{C} = \begin{bmatrix} 1 & C \\ C & 1 \end{bmatrix}, \mathbf{K} = \begin{bmatrix} K_{11} & K_{12} \\ K_{21} & K_{22} \end{bmatrix}, \mathbf{B} = \begin{bmatrix} \beta_1 & 0 \\ 0 & \beta_2 \end{bmatrix}. \tag{1.117}$$

Upon multiplying Equation 1.116 by $\mathbf{C}^{-1}$, we obtain the coupled mode equation

$$\frac{d\mathbf{a}}{dz} = -j\mathbf{Ma}, \tag{1.118}$$

where $\mathbf{a}(z) = [a_1(z),\, a_2(z)]^T$ and $\mathbf{M} = \mathbf{C}^{-1}(\mathbf{K} + \mathbf{BC})$. The elements of the coupling matrix $\mathbf{M}$ are

$$\mathbf{M} = \begin{bmatrix} \gamma_1 & k_{12} \\ k_{21} & \gamma_2 \end{bmatrix}, \tag{1.119}$$

with

$$\gamma_1 = \beta_1 + (K_{11} - CK_{21})/\Delta_C, \tag{1.120}$$

$$\gamma_2 = \beta_2 + (K_{22} - CK_{12})/\Delta_C, \tag{1.121}$$

$$k_{12} = (K_{12} - CK_{22})/\Delta_C, \tag{1.122}$$

$$k_{21} = (K_{21} - CK_{11})/\Delta_C, \tag{1.123}$$

and $\Delta_C = 1 - C^2$. The constants γ_1 and γ_2 are the self-coupling or phase constant of each waveguide, while k_{12} and k_{21} give the mutual coupling strengths between waveguide 1 and waveguide 2. If the two waveguides are identical, then $\beta_1 = \beta_2$, $K_{11} = K_{22}$, and $K_{12} = K_{21}$. For this symmetric case we have $\gamma_1 = \gamma_2$ and $k_{12} = k_{21} = k_c$.

Figure 1.15a and b show the coupling strength k_c as a function of the coupling gap between two identical SOI waveguides for the quasi-TE and quasi-TM modes. The waveguide thickness is 250 nm and three different values of the waveguide width are considered: 300, 400, and 500 nm. The plots show that the coupling strength decreases exponentially with increasing coupling gap. Also note that the coupling strength decreases with increasing waveguide width. This is due to the fact that the modes are more tightly confined in the cores and thus have weaker interaction with each other.

1.3.2   Solution of the coupled mode equations

To solve the coupled mode equation (1.118) for a pair of coupled waveguides, we diagonalize the coupling matrix $\mathbf{M}$ in the form

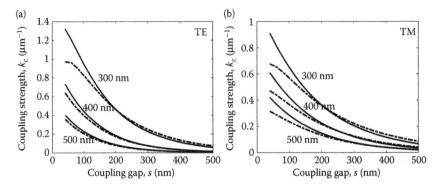

Figure 1.15 Coupling strength k_c as a function of the coupling gap between two SOI waveguides for waveguide widths of 300, 400, and 500 nm: (a) TE polarization and (b) TM polarization. The waveguide thickness is fixed at 250 nm. Solid lines are results obtained using semi-vectorial field distributions of the modes; dashed lines are results obtained using full-vectorial field distributions.

$\mathbf{M} = \mathbf{Q}\,\Lambda\,\mathbf{Q}^{-1}$, where Λ is a diagonal matrix containing the eigenvalues of $\mathbf{M}$ and $\mathbf{Q}$ is a matrix containing its eigenvectors. The solution to Equation 1.118 can then be expressed as

$$\mathbf{a}(z) = \mathbf{Q}e^{-j\Lambda z}\mathbf{Q}^{-1}\mathbf{a}(0) = \mathbf{T}(z)\mathbf{a}(0), \tag{1.124}$$

where $\mathbf{a}(0)$ represents the input fields $[a_1(0), a_2(0)]^T$ to the coupler at $z = 0$, and $\mathbf{T}(z)$ is the transfer matrix of the coupler. The eigenvalues of the matrix $\mathbf{M}$ in Equation 1.119 are found to be

$$\lambda = \gamma_0 \pm \Omega, \tag{1.125}$$

where

$$\Omega = \sqrt{\delta^2 + k_{12}k_{21}}. \tag{1.126}$$

In the above expressions, $\gamma_0 = (\gamma_1 + \gamma_2)/2$ is the average propagation constant and $\delta = (\gamma_2 - \gamma_1)/2$ expresses the degree of asynchronism or phase mismatch between the two waveguides. The eigenvector matrix is

$$\mathbf{Q} = \begin{bmatrix} k_{12} & -(\delta + \Omega) \\ \delta + \Omega & k_{21} \end{bmatrix}. \tag{1.127}$$

Substituting the eigenvalues and eigenvector matrix into Equation 1.124, we obtain the following expressions for the elements of the transfer matrix **T**:

$$T_{11} = T_{22} = \left[\cos(\Omega z) + j\frac{\delta}{\Omega}\sin(\Omega z) \right] e^{-j\gamma_0 z}, \tag{1.128}$$

$$T_{12} = -j\frac{k_{12}}{\Omega}\sin(\Omega z)e^{-j\gamma_0 z}, \tag{1.129}$$

$$T_{21} = -j\frac{k_{21}}{\Omega}\sin(\Omega z)e^{-j\gamma_0 z}. \tag{1.130}$$

For an asymmetric coupler, $\delta^2 > 0$, Equation 1.126 shows that k_{12} and k_{21} are smaller than Ω so that $|T_{21}|^2 < 1$ and $|T_{12}|^2 < 1$. Thus power from one waveguide can never be completely transferred to the other. When the two waveguides are identical, $\delta = 0$ and $\Omega = k_{12} = k_{21} = k_c$; in this case, complete power transfer between the two waveguides is achieved at the coupling length $L_c = \pi/2k_c$.

For a symmetric coupler of length L, the transfer matrix is given by

$$\mathbf{T} = \begin{bmatrix} \cos(k_c L) & -j\sin(k_c L) \\ -j\sin(k_c L) & \cos(k_c L) \end{bmatrix} e^{-j\gamma_0 L}. \tag{1.131}$$

In the analysis of coupled waveguide devices, it is often convenient to model the coupler by a lumped coupling junction connected by two uncoupled waveguides with length L and propagation constant γ_0, as shown in Figure 1.16. The transfer matrix of the lumped coupling junction is given by

$$\mathbf{T} = \begin{bmatrix} \tau & -j\kappa \\ -j\kappa & \tau \end{bmatrix}, \tag{1.132}$$

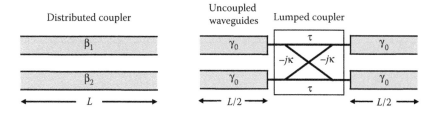

Figure 1.16 Representation of a pair of evanescently coupled waveguides of length L by a lumped coupler connected by two uncoupled waveguides.

where $\tau = \cos(k_cL)$ and $\kappa = \sin(k_cL)$ are the field transmission coefficient and field coupling coefficient, respectively, of the coupling junction. Power conservation requires that $\tau^2 + \kappa^2 = 1$. Note that in an asymmetric coupler, the transmission coefficient has a small phase shift as indicated by Equation 1.128.

When a straight waveguide is coupled to a curved waveguide, the analysis is more complicated since the phase front tilt of the bent waveguide mode must be taken into account. To simplify the problem, we may approximate the curved waveguide by a series of waveguide segments that are parallel to the straight waveguide, as depicted in Figure 1.17b. This approximation is reasonable since the coupling strength between two waveguides decreases exponentially with the gap separation between them, so that the interaction between a straight and bent waveguide quickly drops off as the mismatch between their phase fronts becomes more significant. The above simplification allows us to apply the result for the straight waveguide coupler to compute the coupling coefficient between a curved waveguide and a straight waveguide, or between two curved waveguides. The approach, however, neglects the skewed mode distribution of the bent waveguide and assumes that it can be approximated by that of a straight waveguide with the same width.

Consider the coupling between a straight waveguide and a bent waveguide with inner radius R_1 and outer radius R_2, as shown in Figure 1.17. We denote the edge-to-edge gap separation between the two structures by the function

$$s(\theta) = s_0 + R_2(1 - \cos\theta), \tag{1.133}$$

(a)   (b)

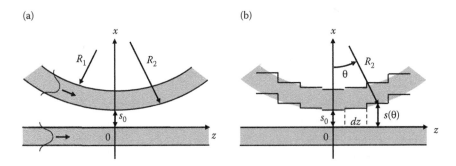

Figure 1.17 (a) Evanescent coupling between a straight waveguide and a bent waveguide. (b) Approximation of the bent waveguide by a series of waveguide segments parallel to the straight waveguide.

with the minimum separation s_0 occurring at $\theta = 0$. Over each segment dz, the field coupling between the two parallel waveguide segments is given by

$$d\kappa(z) = \sin(k_c(s)dz) \approx k_c(s)dz, \tag{1.134}$$

where $k_c(s)$ is the coupling strength between two straight waveguides separated by a gap $s(\theta)$. From Figure 1.15, we find that k_c typically has an exponential dependence on s of the form

$$k_c(s) = k_c(0)e^{-s/\sigma}. \tag{1.135}$$

Since $z = R_2 \sin \theta$ and $dz = R_2 \cos d\theta$, we integrate Equation 1.134 over θ from $-\pi/2$ to $\pi/2$ to get the total field coupling between the straight and bent waveguides:

$$\kappa = \int_{-\pi/2}^{\pi/2} k_c(s)R_2 \cos\theta d\theta. \tag{1.136}$$

The above expression can also be used to estimate the coupling coefficient between two microring waveguides. The only change is in the expression for the gap separation $s(\theta)$. For two microrings with the same outer radii R_2, $s(\theta)$ is given by

$$s(\theta) = s_0 + 2R_2(1 - \cos\theta), \tag{1.137}$$

where s_0 is the minimum gap separation between the two microrings.

The plots in Figure 1.18a and b show the power coupling coefficient κ^2 as a function of the minimum coupling gap s_0 between an SOI microring and a straight waveguide, and between two SOI microrings of the same radius. All waveguides have core dimensions of 250 nm thickness and 300 nm width. The four sets of curves in the plots correspond to microring radius $R = 2, 5, 10,$ and 20 μm for both TE (solid lines) and TM polarization (dashed lines). The plots show that the coupling coefficient decreases rapidly with increasing coupling gap. For the same gap, the coupling is also smaller for smaller microring radius since the interaction length is shorter. Comparison between Figure 1.18a and b also shows that coupling between two microrings is weaker than coupling between a microring and a straight waveguide because the separation distance $s(\theta)$ between the two ring waveguides increases faster in the former case.

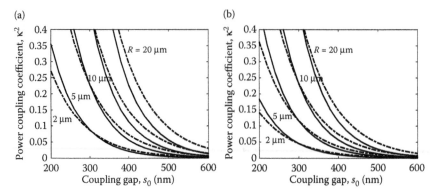

Figure 1.18 Plots of the power coupling coefficient κ^2 vs. the minimum coupling gap s_0 between (a) a microring of radius R and a straight waveguide, and (b) between two microrings of the same radius R. Solid lines are for TE polarization; dashed lines are for TM polarization.

1.4 Fabrication of Microring Resonators

Ring resonators were first realized in the GeO_2-doped silica material (Kominato et al. 1992, Suzuki et al. 1992). Due to the relatively low-index contrast of the material system ($\Delta n < 1\%$), early ring resonators were fabricated with very large radii ($R > 1$ mm) in order to reduce the bending loss. Much more compact microring resonators were subsequently demonstrated using waveguides with high lateral index contrast such as AlGaAs/GaAs (Rafizadeh et al. 1997, Absil et al. 2000) and polysilicon (Little et al. 1998). However, since the high-index contrast significantly reduces the lateral coupling strength, very narrow coupling gaps (~200 nm) between the microring and the bus waveguides were required. Such small coupling gaps were difficult to fabricate using conventional photolithography, so electron beam lithography was typically required to achieve good device performance (Rafizadeh et al. 1997, Absil et al. 2000). One way to avoid the need for patterning very small coupling gaps is to use a vertical coupling configuration which had been demonstrated earlier. (Suzuki et al. 1992). In a vertically coupled microring device, the microring and the bus waveguides were formed in two different high-index layers separated by a thin, low-index coupling layer. Since the layers were either epitaxially grown, as in the case of III–V semiconductor devices (Absil et al. 2001, Grover et al. 2001), or deposited using a chemical vapor deposition process, as in the case of glass-based devices (Little et al. 1999, 2004), their thicknesses could be accurately controlled, which allows for better control of the coupling between the microring and bus waveguides. Another advantage of the vertical coupling configuration is that since the microring resonators and the bus waveguides reside on different layers, the material layer for the microrings could be separately designed for the application of interest, for example, for lasing, photodetection, or electrooptic modulation applications. On the other hand, the fabrication of vertically coupled microring devices involves more processing steps than laterally coupled devices, which adversely impacts device quality and fabrication yield. With electron beam lithography and deep UV lithography (Bogaerts et al. 2005) becoming more economically accessible, most microring devices nowadays are fabricated based on the lateral coupling configuration with lateral feature sizes accurately defined down to 100 nm.

Figure 1.19 depicts a typical process flow for fabricating integrated optics devices including microring resonators using electron beam or UV lithography. The process starts with a substrate consisting of a high-index core layer (e.g., Si) residing on a

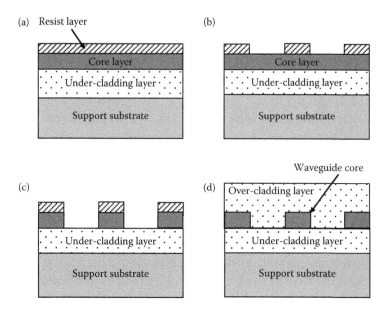

Figure 1.19   Overview of the processing steps for fabricating integrated optics devices: (a) spin-coat a resist layer on a waveguide substrate, (b) expose device pattern and develop resist, (c) transfer device pattern onto core layer by dry-etching, and (d) remove resist and deposit over-cladding layer.

low-index buffer layer (e.g., SiO_2). These layers are either epitaxially grown (e.g., for III–V semiconductors), defined by ion implantation (e.g., for SOI material), or deposited by chemical vapor deposition (for glass-based devices). To define the microring resonators and other device structures on the chip, a photoresist or electron beam resist layer is first spin-coated on the wafer. The device pattern on a mask is then imprinted onto the resist layer by electron beam or UV exposure. Next, the resist is developed and the device pattern is transferred onto the waveguide core layer using a dry etching process such as reactive ion etching (RIE) or inductively coupled plasma (ICP) etching. The resist layer is then washed away and the wafer is coated with an over-cladding layer. Many applications also require additional processing steps to fabricate micro-heaters or electrodes and contact pads on the chip.

Due to random variations in the fabrication process, the fabricated devices typically have dimensions that are slightly deviated from the designed values. As a result, the measured response of the device will in general be different from the target response. One of the most pronounced effects of fabrication-induced variations is the detuning of the resonant frequencies of a microring from the designed values. From Equation 1.70 we find that the change in the resonant wavelength due to a change in the average radius of a microring is given by $\Delta\lambda = (\lambda/R)\Delta R$. As an example, for a 10-µm radius microring resonator operating around the 1.55 µm wavelength, a 1-nm change in the average radius can induce a shift in the resonant wavelengths by as much as 0.15 nm, or almost 20 GHz. Since it is not possible to control fabrication variations to within 1 nm with the current fabrication technology, a post-fabrication method is typically required to correct for the resonance detuning of the microring. A common approach is to use a micro-heater fabricated directly above the microring to thermo-optically tune its resonant frequencies. Several methods for permanently trimming a microring resonance have also been developed. These include UV trimming of the refractive index of the cladding material (Chu et al. 1999), or in the case of silicon waveguides, permanently altering the index of the silicon core by oxidation (Chen et al. 2011, Shen et al. 2011), amorphization (Bachman et al. 2013) or by inducing strain and stress in the material (Schrauwen et al. 2008).

Another practical issue that must be considered in the design of a microring device is polarization dependence. Due to birefringence, the resonant frequencies of TE and TM polarized light in the microring generally do not coincide with each other. In addition, the coupling coefficient between the microring and a bus waveguide, or between two microrings, is also different for the TE and TM polarizations, as evident from Figure 1.18. This polarization sensitivity tends to be more pronounced for microrings made of high-index contrast materials. In general, it is not possible to design the microring waveguide dimensions to completely eliminate the polarization dependence in both the resonant frequencies and the coupling coefficients over a given bandwidth. For this reason, to achieve polarization insensitive operation, a polarization diversity scheme must be used, which consists of splitting the input light into two orthogonal polarization components and feeding each component into a separate microring circuit independently optimized for that polarization. The outputs of the two circuits are then combined to form the total response of the device.

A further issue that must be considered in practical microring device applications is the variation in the resonant frequencies due to fluctuations in the ambient temperature. This is caused by the fact that the refractive indices of the waveguide materials are temperature dependent. For applications requiring the resonant frequency of the microring to be stabilized, the chip is typically placed on a thermoelectric cooler and its temperature actively controlled and monitored by a temperature sensor. It is possible to stabilize the device temperature to less than $\pm 0.1°C$ using a feedback control loop.

1.5   Summary

This chapter provides a brief review of the basic concepts in integrated optics that are essential for understanding the operation of a microring device and the various physical parameters influencing its behaviors. In particular, a familiarity with the theory of optical waveguides, the characteristics of whispering gallery modes in bent waveguides, and the CMT describing evanescent wave coupling is necessary for analyzing and designing microring resonator devices in subsequent chapters of the book. With a view on the practical realization and implementation of these devices, we also provided an overview of the standard lithographic process for fabricating microring resonators and other integrated optics components, and discussed several important issues that must be considered in the practical applications of these devices. In Chapter 2, we will develop basic models for analyzing the spectral responses and determining various performance characteristics of a microring resonator. These models will also be used in subsequent chapters for analyzing and designing more advanced microring devices.

References

Absil, P. P., Hryniewicz, H. V., Little, B. E., Johnson, F. G., Ritter, K. J., Ho, P.-T. 2001. Vertically coupled microring resonators using polymer wafer bonding. *IEEE Photonics Technol. Lett.* 13(1): 49–51.

Absil, P. P., Hryniewicz, J. V., Little, B. E., Wilson, R. A., Joneckis, L. G., Ho, P.-T. 2000. Compact microring notch filters. *IEEE Photonics Technol. Lett.* 12(4): 398–400.

Bachman, D., Chen, Z., Fedosejevs, R., Tsui, Y. Y., Van, V. 2013. Permanent fine tuning of silicon microring devices by femtosecond laser surface amorphization and ablation. *Opt. Express* 21(9): 11048–11056.

Berglund, W., Gopinath, A. 2000. WKB analysis of bend losses in optical waveguides. *J. Lightwave Technol.* 18(8): 1161–1166.

Bogaerts, W., Baets, R., Dumon, P., Wiaux, V., Beckx, S., Taillaert, D., Luyssaert, B., Van Campenhout, J., Bienstman, P., Van Thourhout, D. 2005. Nanophotonic waveguides in silicon-on-insulator fabricated with CMOS technology. *J. Lightwave Technol.* 23(1): 401–412.

Chen, C. J., Zheng, J., Gu, T., McMillan, J. F., Yu, M., Lo, G. Q., Kwong, D. L., Wong, C. W. 2011. Selective tuning of high-Q silicon photonic crystal nanocavities via laser-assisted local oxidation. *Opt. Express* 19(13): 12480–12489.

Cheng, Y., Lin, W., Fujii, Y. 1990. Local field analysis of bent graded-index planar waveguides. *J. Lightwave Technol.* 8(10): 1461–1469.

Chiang, K. S. 1993. Review of numerical and approximate methods for the modal analysis of general optical dielectric waveguides. *Opt. Quantum Electron.* 26: S113–S134.

Chiang, K. S. 1996. Analysis of the Effective Index Method for the vector modes of rectangular-core dielectric waveguides. *IEEE Trans. Microwave Theory Tech.* 44(5): 692–700.

Chin, M. K., Ho, S. T. 1998. Design and modeling of waveguide-coupled single-mode microring resonators. *J. Lightwave Technol.* 16(8): 1433–1446.

Chu, S. T., Pan, W., Sato, S., Taneko, T., Little, B. E., Kokubun, Y. 1999. Wavelength trimming of a microring resonator filter by means of a UV sensitive polymer overlay. *IEEE Photonics Technol. Lett.* 11(6): 688–690.

Chuang, S.-L. 1987a. A coupled mode formulation by reciprocity and a variational principle. *J. Lightwave Technol.* 5(1): 5–15.

Chuang, S.-L. 1987b. A coupled-mode theory for multiwaveguide systems satisfying the reciprocity theorem and power conservation. *J. Lightwave Technol.* 5(1): 174–183.

Coale, F. S. 1956. A traveling-wave directional filter. *IRE Trans. Microwave Theory. Tech.* 4(4): 256–160.

Feng, N.-N., Zhou, G.-R., Xu, C., Huang, W.-P. 2002. Computation of full-vector modes for bending waveguide using cylindrical Perfectly Matched Layers. *J. Lightwave Technol.* 20(11): 1976–1980.

Frey, B. J., Leviton, D. B., Madison, T. J. 2006. Temperature-dependent refractive index of silicon and germanium. In *SPIE Astronomical Telescopes + Instrumentation*, International Society for Optics and Photonics, Orlando, FL, article no. 62732J.

Grover, R., Absil, P. P., Van, V., Hryniewicz, J. V., Little, B. E., King, O., Calhoun, L. C., Johnson, F. G., Ho, P.-T. 2001. Vertically coupled GaInAsP-InP microring resonators. *Opt. Lett.* 26(8): 506–508.

Haus, H. A., Huang, W. P., Kawakami, S., Whitaker, N. A. 1987. Coupled mode theory of optical waveguides. *J. Lightwave Technol.* 5: 16–23.

Heiblum, M., Harris, J. H. 1975. Analysis of curved optical waveguides by conformal transformation. *IEEE J. Quantum Electron.* QE-11(2): 75–83.

Hiremath, K. R., Hammer, M., Stoffer, R., Prkna, L., Ctyroky, J. 2005. Analytic approach to dielectric optical bent slab waveguides. *Opt. Quantum Electron.* 37(1–3): 37–61.

Jackson, J. D. 1999. *Classical Electrodynamics*, 3rd Ed. Hoboken, NJ: John Wiley & Sons.

Kakihara, K., Kono, N., Saitoh, K., Koshiba, M. 2006. Full-vectorial finite element method in a cylindrical coordinate system for loss analysis of photonic wire bends. *Opt. Express* 14(23): 11128–11141.

Knox, R. M., Toulios, P. P. 1970. Integrated circuits for the millimeter through optical frequency range. In *Symposium on Submillimeter Waves*, New York, pp. 497–516.

Koshiba, M. 1992. *Waveguide Theory by the Finite Element Method*. London: Kluwer Academic Publishers.

Kominato, T., Ohmori, Y., Takato, N., Okazaki, H., Yasu, M. 1992. Ring resonators composed of GeO_2-doped silica waveguides. *J. Lightwave Technol.* 10(12): 1781–1788.

Little, B. E., Chu, S. T., Absil, P. P., Hryniewicz, J. V., Johnson, F. G., Seiferth, F., Gill, D., Van, V., King, O., Trakalo, M. 2004. Very high-order microring resonator filters for WDM applications. *IEEE Photonics Technol. Lett.* 16(10), 2263–2265.

Little, B. E., Chu, S. T., Pan, W., Ripin, D., Kaneko, T., Kokubun, Y., Ippen, E. 1999. Vertically coupled glass microring resonator channel dropping filters. *IEEE Photonics Technol. Lett.* 11(2): 215–217.

Little, B. E., Foresi, J. S., Steinmeyer, G., Thoen, E. R., Chu, S. T., Haus, H. A., Ippen, E. P., Kimerling, L. C., Greene, W. 1998. Ultra-compact Si-SiO_2 microring resonator optical channel dropping filters. *IEEE Photonics Technol. Lett.* 10(4): 549–551.

Lui, W. W., Xu, C.-L. Xu, Hirono, T., Yokoyama, K., Huang, W.-P. 1998. Full-vectorial wave propagation in semiconductor optical bending waveguides and equivalent straight waveguide approximations. *J. Lightwave Technol.* 16(5): 910–914.

Malitson, I. H. 1965. Interspecimen comparison of the refractive index of fused silica. *J. Opt. Soc. Am.* 55:1205–1209.

Marcatili, E. A. 1969a. Dielectric rectangular waveguide and directional coupler for integrated optics. *Bell Syst. Tech. J.* 48:2071–2102.

Marcatili, E. A. 1969b. Bends in optical dielectric guides. *Bell Syst. Tech. J.* 48(7): 2103–2132.

Marcuse, D. 1969. Mode conversion caused by surface imperfections of a dielectric slab waveguide. *Bell Syst. Tech. J.* 48:3187–3215.

Marcuse, D. 1972. *Light Transmission Optics*. New York: Van Nostrand Reinhold.

Okamoto, K. 2000. *Fundamentals of Optical Waveguides*. San Diego: Academic Press.

Rafizadeh, D., Zhang, J. P., Hagness, S. C., Taflove, A., Stair, K. A., Ho, S. T. 1997. Waveguide-coupled AlGaAs/GaAs microcavity ring and disk resonators with high finesse and 21.6-nm free spectral range. *Opt. Lett.* 22(16): 1244–1246.

Rahman, B. M. A., Davies, J. B. 1984. Finite-element solution of integrated optical waveguides. *J. Lightwave Technol.* LT-2(5): 682–688.

Rivera, M. 1995. A finite difference BPM analysis of bent dielectric waveguides. *J. Lightwave Technol.* 13(2): 233–238.

Rowland, D. R., Love, J. D. 1993. Evanescent wave coupling of whispering gallery modes of a dielectric cylinder. *Inst. Elect. Eng. Proc. J.* 140:177–188.

Prabhu, A. M., Tsay, A., Han, Z., Van, V. 2010. Extreme miniaturization of silicon add-drop microring filters for VLSI photonics applications. *IEEE Photonics J.* 2(3): 436–444.

Schrauwen, J., Van Thourhout, D., Baets R. 2008. Trimming of silicon ring resonator by electron beam induced compaction and strain. *Opt. Express* 16(6): 3738–3743.

Shen, Y., Divliansky, I. B., Basov, D. N., Mookherjea, S. 2011. Electric-field-driven nano-oxidation trimming of silicon microrings and interferometers. *Opt. Lett.* 36(14): 2668–2670.

Steinlechner, J., Krüger, C., Lastzka, N., Steinlechner, S., Khalaidovski, A., Schnabel, R. 2013. Optical absorption measurements on crystalline silicon test masses at 1550 nm. *Classical Quantum Gravity* 30(9): 095007.

Suzuki, S., Shuto, K., Hibino, Y. 1992. Integrated-optic ring resonators with two stacked layers of silica waveguide on Si. *IEEE Photonics Technol. Lett.* 4(11): 1256–1258.

Tien, P. K. 1971. Light waves in thin films and integrated optics. *Appl. Opt.* 10: 2395–2419.

Vlasov, Y. A., McNab, S. J. 2004. Losses in single-mode silicon-on-insulator strip waveguides and bends. *Opt. Express* 12: 1622–1631.

Xiao, J., Ma, H., Bai, N., Liu X., Sun, X. 2009. Full-vectorial analysis of bending waveguides using finite difference method based on H-fields in cylindrical coordinate systems. *Opt. Commun.* 282: 2511–2515.

Xu, C. L., Huang, W. P., Stern, M. S., Chaudhuri, S. K. 1994. Full-vectorial mode calculations by finite difference method. *IEE Proc.: Optoelectron.* 141(5): 281–286.

Xu, Q., Fattal, D., Beausoleil, R. G. 2008. 1.5-μm-radius high-Q silicon microring resonators. *Opt. Express* 16: 4309–4315.

Yariv, A. 1973. Coupled-mode theory for guided-wave optics. *IEEE J. Quantum Electron.* 9: 919–933.

CHAPTER 2

Analytical Models of a Microring Resonator

In Section 1.2.2, we showed that as light travels around a microring resonator, the periodic boundary condition requiring the field to be replicated after every round-trip gives rise to discrete longitudinal resonant modes. At a resonant frequency, constructive interference of the field causes a large amount of energy to build up in the microring. In practice, this energy must be supplied to the resonator by an external source and, for useful applications, one must also be able to extract the light out of the microring. In practical implementations of microring devices, the coupling of light into and out of the resonator is accomplished by evanescent wave coupling between the microring and one or two external bus waveguides, as depicted in Figure 1.1. A microring coupled to a single bus waveguide is commonly referred to as the all-pass configuration, whereas a microring coupled to two bus waveguides is called the add-drop configuration.

The spectral characteristics of a microring resonator are very similar to those of a Fabry–Perot (FP) resonator, even though the former is a traveling-wave resonator, while the latter is a standing-wave resonator. Specifically, the all-pass microring behaves like an FP resonator with one partially reflecting mirror, while the add-drop microring behaves like one with both partially reflecting mirrors. The distinction between a standing-wave resonator and a traveling-wave resonator is that the field at any point in the former structure is the result of the interference between two waves traveling in opposite directions, whereas the field inside a traveling-wave resonator is that of either the forward- or the backward-traveling wave alone. As a result, the field in a standing-wave resonator has a spatially-dependent amplitude distribution, whereas the field amplitude in a traveling-wave resonator is nearly uniform. An important advantage of microring resonators over FP resonators is that they do not require mirrors, which are generally difficult to realize in integrated optics. In addition, due to the unidirectional propagation of light in the microring, the output light signals of a microring device appear at physically isolated ports from the input light, which is desirable for most integrated optics applications.

In this chapter, we develop the analytical models that will allow us to determine the transmission responses and other spectral characteristics of a microring resonator. In general, there are two alternate formalisms that can be used to analyze the response of a microring device: the power coupling formalism and the energy coupling formalism. The microring models based on these formalisms will also be useful for analyzing and designing more complicated device configurations as well as more advanced applications of microring resonators in later chapters. We will begin in Section 2.1 by examining the resonance characteristics of a microring resonator in the absence of coupling to an external waveguide. In Section 2.2 we will derive the transfer functions of a microring resonator coupled to one and two external waveguides using the power coupling formalism. Alternative descriptions of the all-pass and add-drop microring resonators based on the energy coupling formalism will be given in Section 2.3. Finally, Section 2.4 will establish the relationship between the power coupling and energy coupling formalisms and discuss the validity of each model for analyzing microring devices.

2.1   Resonance Spectrum of an Uncoupled Microring Resonator

Important resonant characteristics of a microring resonator can be determined by considering the simple structure of an isolated microring without any coupling to external waveguides. In Section 1.2.2, we found that the resonant condition for a monochromatic light wave traveling around the microring in either the clockwise or counterclockwise direction is $\beta_\theta = m$, where β_θ is the angular propagation constant and m is the longitudinal resonant mode number. By writing $\beta_\theta = n_r(2\pi/\lambda)R_{\text{eff}}$, where n_r is the effective index of the microring waveguide at the effective radial distance[*] R_{eff} from the center, we can express the resonant condition as[†]

$$2\pi R_{\text{eff}} = m\lambda/n_r. \tag{2.1}$$

---

[*] The effective radius R_{eff} is defined in Equation 1.57.
[†] The transverse resonant mode number appears implicitly in this resonant condition through the effective index n_r and the effective radius R_{eff} of the transverse mode. See Equation 1.70 for an explicit resonant condition containing both the transverse and longitudinal mode numbers.

The above expression states that resonance occurs whenever the round-trip path length of the microring is equal to an integer multiple of the guided wavelength. Approximating the effective radius by the average radius R of the microring, we obtain from Equation 2.1 the free space wavelength λ_m of the resonant mode m:

$$\lambda_m = \frac{2\pi n_r R}{m}. \tag{2.2}$$

The resonance spectrum of a microring resonator thus consists of a comb of resonance peaks occurring at discrete wavelengths λ_m. The spacing between two consecutive resonances is called the free spectral range (FSR) of the resonator. The FSR can be computed by observing that it is equal to the wavelength change required for the round-trip phase of the resonator to undergo a 2π shift. From the expression for the round-trip phase of the microring, $\phi_{rt} = n_r(2\pi/\lambda)2\pi R$, we take the derivative with respect to the wavelength to get

$$\frac{d\phi_{rt}}{d\lambda} = \left(\frac{dn_r}{d\lambda} - \frac{n_r}{\lambda}\right)\frac{2\pi}{\lambda}2\pi R. \tag{2.3}$$

Since the group index of the microring waveguide is given by $n_g = n_r - \lambda dn_r/d\lambda$, we can also write Equation 2.3 as

$$\frac{d\phi_{rt}}{d\lambda} = -n_g \frac{2\pi}{\lambda^2}2\pi R. \tag{2.4}$$

The FSR is defined as the wavelength change $\Delta\lambda_{FSR}$ required to obtain a round-trip phase change $\Delta\phi_{rt} = 2\pi$:

$$\Delta\lambda_{FSR} = \frac{2\pi}{d\phi_{rt}/d\lambda} = -\frac{\lambda^2}{2\pi n_g R}. \tag{2.5}$$

In terms of frequency, the FSR is given by

$$\Delta f_{FSR} = -\frac{c}{\lambda^2}\Delta\lambda_{FSR} = \frac{c}{2\pi n_g R} = \frac{1}{T_{rt}}, \tag{2.6}$$

where $T_{rt} = 2\pi n_g R/c$ is the round-trip group delay (GD), that is, the time it takes a pulse to travel around the microring. Equation 2.6 shows that the FSR is inversely proportional to the microring radius. Note that since the group index of the microring wave-guide also exhibits frequency dispersion, the FSR will also depend on the frequency in general. This effect can be observed in Figure 1.10a and b, which show that the wavelength spacings between successive resonant modes are not constant. The frequency dispersion of the FSR of a microring can have a significant impact in many applications. For example, it places an upper limit on the wavelength conversion bandwidth of the four-wave mixing process in a nonlinear microring resonator. This will be discussed in detail in Chapter 4.

In Section 1.2.2, we also found that the field amplitude of a resonant mode is constant along the direction of propagation (the angular direction), implying that the power distribution is uniform around the microring. From Equation 1.68 we may express the amplitude of the resonant mode m as

$$A(t) = A_0 e^{j\omega_m t} e^{-\gamma_m t}, \tag{2.7}$$

where ω_m is the resonant frequency and γ_m is the decay rate of the amplitude of mode m. The amplitude decay rate is related to the cavity lifetime* by $\gamma_m = 1/2\tau_m$, or to the power attenuation constant α of the microring waveguide through Equation 1.72,

$$\gamma_m = \frac{\omega_m \alpha/2}{m/R_{\text{eff}}} = \frac{c\alpha}{2n_r}. \tag{2.8}$$

The amplitude A_0 in Equation 2.7 is usually normalized with respect to either energy or power. In the energy normalization, A_0 is normalized such that $|A_0|^2$ gives the initial stored energy in the microring and $|A(t)|^2$ represents the stored energy at time t. In the power normalization, $|A_0|^2$ and $|A(t)|^2$ give the power flow through any cross section of the microring waveguide at the initial time and at time t, respectively.

To obtain the frequency response of the mode amplitude, we take the Fourier transform of Equation 2.7 to get

$$A(\omega) = \frac{A_0}{j(\omega - \omega_m) + \gamma_m}. \tag{2.9}$$

---

* Recall that we define the cavity lifetime (or photon lifetime) τ_m as the time it takes the energy of mode m to decay to $1/e$ of its initial value.

The energy or power spectrum is obtained by taking the absolute square of the above expression,

$$|A(\omega)|^2 = \frac{|A_0|^2}{(\omega - \omega_m)^2 + \gamma_m^2}. \qquad (2.10)$$

The above equation shows that the energy or power spectrum of the microring has the form of a Lorentzian resonance centered around the resonant frequency ω_m. The full width at half max (FWHM) bandwidth, or linewidth of the resonance, is obtained by equating the expression in Equation 2.10 to half its peak value. The result is

$$\Delta\omega_{BW} = 2\gamma_m = \frac{\omega_m}{Q_0}, \qquad (2.11)$$

where $Q_0 = \omega_m \tau_m$ is the intrinsic Q factor of mode m given in Equation 1.77. The above relation shows that the resonance linewidth is inversely proportional to the Q factor (or to the cavity lifetime). It also gives the well-known formula for calculating the Q factor as the ratio of the resonant frequency to the FWHM bandwidth, $Q_0 = \omega_m/\Delta\omega_{BW} = \lambda_m/\Delta\lambda_{BW}$. Another useful formula relating the intrinsic Q factor to the power attenuation coefficient α of the microring waveguide is obtained by using Equation 2.8 for γ_m:

$$Q_0 = \frac{2\pi n_r}{\alpha\lambda_m}. \qquad (2.12)$$

2.2 Power Coupling Description of a Microring Resonator

In this section, we determine the spectral response of a microring resonator in the presence of coupling to one or two external bus waveguides. There are two different approaches which we can use to analyze such a structure: the power coupling approach (or "coupling of modes in space") and the energy coupling approach (or "coupling of modes in time"). In the power coupling approach, the device response is described in terms of power exchange between the microring and the external bus waveguides. Couplings between the microring and the bus waveguides are described by the field coupling coefficients, which are obtained from the solution of the coupled mode equations in space (Section 1.3.2). In the energy coupling approach, the behavior of the device is described from the point of view of energy exchange between the microring and the external bus waveguides. Strictly speaking, the energy coupling formalism is valid only for weakly coupled microring resonators, whereas there is no restriction on the coupling strengths in the power coupling approach. In this section we will use the power coupling approach to derive the transfer functions of a microring resonator in the add-drop and all-pass configurations. Analysis of these structures based on the energy coupling formalism will be given in Section 2.3.

2.2.1   The add-drop microring resonator

An add-drop microring resonator (ADMR) is a four-port optical device which consists of a microring evanescently coupled to an input waveguide and an output waveguide, as shown in Figure 2.1a. The four ports are labeled input, through, drop, and add. An optical signal applied to the input port with frequency coinciding to a microring resonance will be transmitted (or "dropped") at the drop port. If the frequency of the input signal is tuned off resonance, the light will instead be transmitted to the through port. The device can thus be used as an add-drop filter in a wavelength division multiplexing (WDM) communication network. It allows a wavelength channel from an input WDM signal that is in resonance with the microring to be extracted to the drop port while allowing all other

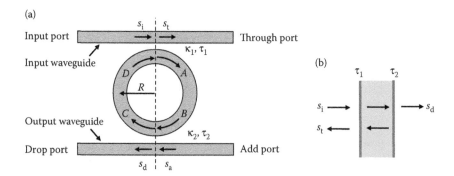

Figure 2.1   (a) Schematic of an ADMR with average radius R, input coupling coefficient κ_1 and output coupling coefficient κ_2. (b) In terms of transfer characteristics, the ADMR is equivalent to an FP resonator with mirror reflectivities τ_1 and τ_2.

channels to bypass to the through port. Conversely, a channel that is in resonance with the microring can be added onto the incoming WDM stream via the add port.

To derive the transfer functions at the drop port and through port of an ADMR, we consider the microring in Figure 2.1a with average radius R evanescently coupled to an input and an output waveguide. We represent each coupling junction between the microring and a bus waveguide as a lumped coupler with field coupling coefficient κ_i and transmission coefficient τ_i, where $i = 1$ for the input and $i = 2$ for the output. For simplicity we assume the coupling junctions to be lossless, so that $\kappa_i^2 + \tau_i^2 = 1$.* Suppose a monochromatic light wave of frequency ω is applied to the input port. Its electric field can be expressed as

$$\mathcal{E}_i(\mathbf{r},t) = E_0 e(x,y) e^{j(\omega t - \beta_b z)}, \tag{2.13}$$

where $e(x, y)$ is the modal field distribution normalized to unit power, β_b is the propagation constant of the bus waveguide, and z is the direction of propagation along the input waveguide. Let s_i represent the complex amplitude of the input signal at a point z_0 just before the input coupling junction, $s_i = E_0 e^{-j\beta_b z_0}$. Note that s_i is normalized such that the power in the input waveguide is given by $|s_i|^2 = |E_0|^2$. In a similar manner we denote s_a as the amplitude of the light wave applied to the add port, and s_t and s_d as the output signals at the through port and drop port, respectively.

At the input and output coupling junctions, the applied signals at the input and add ports are coupled into the microring, giving rise to a clockwise-propagating wave in the resonator. We label the complex amplitude of this wave at the points just before and after the input and output coupling junctions as A, B, C, and D, as shown in Figure 2.1a. Using the transfer matrix in Equation 1.132 for the input coupler, we have

$$s_t = -j\kappa_1 D + \tau_1 s_i, \tag{2.14}$$

$$A = -j\kappa_1 s_i + \tau_1 D. \tag{2.15}$$

---

* To account for coupling junction loss, we replace the coupling and transmission coefficients by $\sqrt{\alpha_i}\,\kappa_i$ and $\sqrt{\alpha_i}\,\tau_i$, respectively, such that $\alpha_i \kappa_i^2 + \alpha_i \tau_i^2 = \alpha_i$, where α_i is the power attenuation due to the junction loss of coupler i.

Assuming that there is no applied signal at the add port, $s_a = 0$, we also obtain the following relations at the output coupling junction:

$$s_d = -j\kappa_2 B,$$ (2.16)

$$C = \tau_2 B.$$ (2.17)

The fields B and D can be related to the fields A and C, respectively, by

$$B = Ae^{-j\gamma\pi R} = A\sqrt{a_{rt}}\,e^{-j\phi_{rt}/2},$$ (2.18)

$$D = Ce^{-j\gamma\pi R} = C\sqrt{a_{rt}}\,e^{-j\phi_{rt}/2},$$ (2.19)

where $\gamma = \beta - j\alpha/2$ is the complex propagation constant of the microring waveguide, $\phi_{rt} = \beta 2\pi R$ is the round-trip phase, and $a_{rt} = e^{-\alpha\pi R}$ is the round-trip field attenuation factor in the microring. The propagation loss constant α accounts for the total waveguide loss, including absorption, bending loss, and scattering. Using Equations 2.18 and 2.19 to eliminate C and B in Equation 2.17, we get

$$D = \tau_2 A a_{rt} e^{-j\phi_{rt}}.$$ (2.20)

Substituting the above result into Equation 2.15 and solving for A, we obtain

$$A = \frac{-j\kappa_1 s_i}{1 - \tau_1 \tau_2 a_{rt} e^{-j\phi_{rt}}}.$$ (2.21)

Similar expressions for B, C, and D can also be derived in terms of s_i using Equations 2.17 through 2.19. These signals are approximately equal to A in magnitude if the microring loss and coupling coefficients are small ($a_{rt} \approx 1$, $\tau_1 \approx 1$, $\tau_2 \approx 1$). For these high-Q resonators, the field amplitude can be assumed to be uniformly distributed around the microring.[*] However, for strongly coupled or high-loss microring resonators, the field amplitude distribution will generally be nonuniform.

---

[*] This is in contrast to a FP resonator, where the field amplitude distribution is determined by the standing-wave pattern in the cavity.

We note from Equation 2.21 that the field amplitude $|A|$ is maximum at the resonant wavelengths λ_m, where the round-trip phase $\phi_{rt} = 2m\pi$. We can define the field enhancement factor, FE, as the ratio of the peak field amplitude inside the microring at resonance to the input field amplitude:

$$FE = \frac{|A|}{|s_i|}\bigg|_{\phi_{rt}=2m\pi} = \frac{\kappa_1}{1-\tau_1\tau_2 a_{rt}}. \tag{2.22}$$

If the microring is symmetrically coupled ($\tau_1 = \tau_2$) and has low loss ($a_{rt} \approx 1$), the above equation gives $FE \approx 1/\kappa$, that is, the field enhancement factor is inversely proportional to the field coupling coefficient. Thus, by choosing a small coupling coefficient ($\kappa \sim 0.1\text{–}0.01$), we can amplify the field circulating in the microring by 1–2 orders of magnitude with respect to the input light wave. This resonance field enhancement has been widely exploited to reduce the power requirement in nonlinear optics applications of microring resonators, such as all-optical switching and frequency conversion. We will discuss some of these applications in Chapter 4.

To determine the transfer function at the drop port of the ADMR, we substitute $B = A\sqrt{a_{rt}}e^{-j\phi_{rt}/2}$ into Equation 2.16 and, using Equation 2.21, obtain the following result:

$$H_d(\phi_{rt}) \equiv \frac{s_d}{s_i} = -\frac{\kappa_1\kappa_2\sqrt{a_{rt}}e^{-j\phi_{rt}/2}}{1-\tau_1\tau_2 a_{rt}e^{-j\phi_{rt}}}. \tag{2.23}$$

Similarly, substituting $D = \tau_2 A a_{rt} e^{-j\phi_{rt}}$ into Equation 2.14 and using Equation 2.21, we obtain the transfer function at the through port as

$$H_t(\phi_{rt}) \equiv \frac{s_t}{s_i} = \frac{\tau_1 - \tau_2 a_{rt}e^{-j\phi_{rt}}}{1-\tau_1\tau_2 a_{rt}e^{-j\phi_{rt}}}. \tag{2.24}$$

In arriving at the above result, we have also made use of the relation $\tau_1^2 + \kappa_1^2 = 1$. If a signal s_a is applied to the add port of the ADMR and there is no input signal ($s_i = 0$), the transfer functions at the through port (s_t/s_a) and drop port (s_d/s_a) are also given by Equations 2.23 and 2.24, respectively, with the coefficients τ_1 (κ_1) and τ_2 (κ_2) interchanged.

The ADMR can be regarded as an infinite impulse response (IIR) digital filter (Madsen and Zhao 1999, Chapter 3) since the output signal at the drop port of the microring can be expressed as an infinite series of signals, with each successive signal delayed by one round-trip time. Indeed, by defining the round-trip delay variable $z^{-1} = e^{-j\phi_{rt}}$, we can express the transfer functions in Equations 2.23 and 2.24 as

$$H_d(z^{-1}) = -\frac{\kappa_1\kappa_2\sqrt{a_{rt}}\,z^{-1/2}}{1 - \tau_1\tau_2 a_{rt} z^{-1}}, \tag{2.25}$$

$$H_t(z^{-1}) = \frac{\tau_1 - \tau_2 a_{rt} z^{-1}}{1 - \tau_1\tau_2 a_{rt} z^{-1}}. \tag{2.26}$$

The above expressions have the form of a transfer function of a digital filter with unit delay variable z^{-1}. Note that both are first-order transfer functions with a pole at $z^{-1} = 1/\tau_1\tau_2 a_{rt}$. Additionally, the through port transfer function has a zero at $z^{-1} = \tau_1/\tau_2 a_{rt}$, while the drop port transfer function has a trivial zero at $z^{-1} = 0$. By taking the inverse z-transforms of H_d and H_t, we can obtain the discrete-time impulse responses of the ADMR at the drop port and through port, respectively.

The power spectral responses at the drop port and through port of the ADMR can be obtained by taking the absolute square of Equations 2.23 and 2.24:

$$T_d = \left|\frac{s_d}{s_i}\right|^2 = \frac{\kappa_1^2\kappa_2^2 a_{rt}}{1 + \tau_1^2\tau_2^2 a_{rt}^2 - 2\tau_1\tau_2 a_{rt}\cos\phi_{rt}}, \tag{2.27}$$

$$T_t = \left|\frac{s_t}{s_i}\right|^2 = \frac{\tau_1^2 + \tau_2^2 a_{rt}^2 - 2\tau_1\tau_2 a_{rt}\cos\phi_{rt}}{1 + \tau_1^2\tau_2^2 a_{rt}^2 - 2\tau_1\tau_2 a_{rt}\cos\phi_{rt}}. \tag{2.28}$$

Figure 2.2 shows representative plots of the spectral responses at the drop port and through port of an ADMR as functions of the round-trip phase ϕ_{rt}. We see that the drop port response has transmission peaks at the resonances, while the through port response exhibits corresponding transmission dips. By substituting $\phi_{rt} = 2m\pi$ into Equations 2.27 and 2.28, we obtain the values for the maximum transmission at the drop port, $T_{d,max}$, and the minimum transmission at the through port, $T_{t,min}$, as follows:

$$T_{d,max} = \frac{\kappa_1^2\kappa_2^2 a_{rt}}{1 + \tau_1^2\tau_2^2 a_{rt}^2 - 2\tau_1\tau_2 a_{rt}} = \frac{\kappa_1^2\kappa_2^2 a_{rt}}{(1 - \tau_1\tau_2 a_{rt})^2}, \tag{2.29}$$

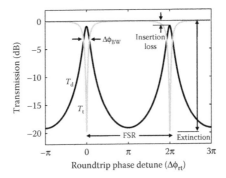

Figure 2.2 Power spectral responses at the drop port (black line) and through port (gray line) of an ADMR with power coupling coefficients $\kappa_1^2 = \kappa_2^2 = 0.2$ and 5% round-trip power loss ($a_{rt} = 0.975$).

$$T_{t,min} = \frac{\tau_1^2 + \tau_2^2 a_{rt}^2 - 2\tau_1\tau_2 a_{rt}}{1 + \tau_1^2 \tau_2^2 a_{rt}^2 - 2\tau_1\tau_2 a_{rt}} = \left(\frac{\tau_1 - \tau_2 a_{rt}}{1 - \tau_1\tau_2 a_{rt}}\right)^2. \tag{2.30}$$

Using the above expressions, we can express the power spectral responses in Equations 2.27 and 2.28 as

$$T_d(\phi_{rt}) = \frac{T_{d,max}}{1 + F\sin^2(\phi_{rt}/2)}, \tag{2.31}$$

$$T_t(\phi_{rt}) = \frac{T_{t,min} + F\sin^2(\phi_{rt}/2)}{1 + F\sin^2(\phi_{rt}/2)}, \tag{2.32}$$

$$F = \frac{4\tau_1\tau_2 a_{rt}}{(1 - \tau_1\tau_2 a_{rt})^2}. \tag{2.33}$$

The parameter F is called the contrast of the resonator and provides a measure of the sharpness of the resonance peak. Equation 2.31 shows that the drop port response of an ADMR has the form of an Airy function, which is characteristic of a FP resonance. Indeed, the drop port and through port transfer functions of an ADMR are formally identical to those of an FP resonator with mirror reflectivities τ_1 and τ_2, as shown in Figure 2.1b. Note, however, that in the ADMR the transmitted signal at the through port is physically isolated from the input signal instead of being reflected back into the input port, as in the case of the FP resonator. The fact that there is no back-reflected signal[*] at the input port of an ADMR makes it more attractive than FP resonators for photonic integrated circuit applications.

For a lossless and symmetrically coupled microring resonator ($a_{rt} = 1$, $\tau_1 = \tau_2$), Equations 2.29 and 2.30 give $T_{d,max} = 1$ and $T_{t,min} = 0$. Thus, at the resonant frequencies, all the input power is transmitted to the drop port and complete extinction is achieved at the through port. In this case the field buildup in the microring is such that the fraction of light coupled out onto the input waveguide interferes destructively with the input signal, resulting in zero transmitted power at the through port. Since power must be conserved in the absence of loss, the input power is completely transferred to the drop port. This process is called resonance-assisted power transfer between the input and output waveguides. Note that complete power transfer is possible even if there is phase mismatch between the bus waveguides and the microring waveguide.[†] The only requirement for complete power transfer in a lossless ADMR is the matching of the input and output coupling coefficients.

When loss is present in the microring, the peak drop port transmission is always less than unity. The difference in decibels between the peak power transmission and unity is called the insertion loss of the device. Complete extinction at the through port can still be achieved in the presence of loss by a suitable choice of the coupling coefficients. In particular, we find from Equation 2.30 that $T_{t,min} = 0$ if $\tau_1 = \tau_2 a_{rt}$. This condition, which results in zero transmission at the through port, is called critical coupling in an ADMR.

To determine the FWHM (or 3 dB) bandwidth $\Delta\phi_{BW}$ of the ADMR, we equate the drop port power transmission in Equation 2.31 to half its peak value,

$$T_d(\Delta\phi_{BW}/2) = \frac{T_{d,max}}{1 + F\sin^2(\Delta\phi_{BW}/4)} = \frac{T_{d,max}}{2}, \tag{2.34}$$

which gives

---

[*] In practice, there is a small back-reflected signal at the input port of an ADMR due to backscattering from waveguide imperfections and the coupling junctions; however, this reflected signal can be minimized with proper design and an optimized fabrication process.

[†] Recall that in an evanescent wave coupler, complete power transfer is not possible if there is phase mismatch between the two waveguides, that is, if they have different propagation constants.

$$\sin(\Delta\phi_{BW}/4) = 1/\sqrt{F}. \tag{2.35}$$

Note that the above expression can also be obtained from the through port power transmission in Equation 2.32 by equating $T_t(\Delta\phi_{BW}/2) = (1 + T_{t,min})/2$. For devices with narrow bandwidths, we can use the small-angle approximation of the sine function in Equation 2.35 to get

$$\Delta\phi_{BW} \approx \frac{4}{\sqrt{F}} = \frac{2(1 - \tau_1\tau_2 a_{rt})}{\sqrt{\tau_1\tau_2 a_{rt}}}. \tag{2.36}$$

Since $\Delta\phi_{BW} = \Delta\omega_{BW}T_{rt}$,[*] where T_{rt} is the round-trip GD, we also have the following expressions for the FWHM bandwidths in terms of frequency and wavelength:

$$\Delta f_{BW} = \frac{1}{2\pi}\frac{\Delta\phi_{BW}}{T_{rt}} = \frac{v_g}{\pi^2 R\sqrt{F}} = \frac{v_g}{2\pi^2 R}\frac{1 - \tau_1\tau_2 a_{rt}}{\sqrt{\tau_1\tau_2 a_{rt}}}, \tag{2.37}$$

$$\Delta\lambda_{BW} = \frac{\lambda_m^2}{\pi^2 n_g R\sqrt{F}} = \frac{\lambda_m^2}{2\pi^2 n_g R}\frac{1 - \tau_1\tau_2 a_{rt}}{\sqrt{\tau_1\tau_2 a_{rt}}}. \tag{2.38}$$

In the above equations, v_g and n_g are the group velocity and group index, respectively, of the microring waveguide.

Using the expression for the bandwidth $\Delta\lambda_{BW}$ in Equation 2.38, we can determine the loaded quality factor of the ADMR as[†]

$$Q = \frac{\lambda_m}{\Delta\lambda_{BW}} = \frac{\pi^2 n_g R\sqrt{F}}{\lambda_m} = \frac{2\pi^2 n_g R}{\lambda_m}\frac{\sqrt{\tau_1\tau_2 a_{rt}}}{1 - \tau_1\tau_2 a_{rt}}. \tag{2.39}$$

For low-loss microrings ($a_{rt} \approx 1$) the loaded Q factor is dominated by external couplings to the bus waveguides. If the input and output

---

[*] Since the round-trip phase of the microring is $\phi_{rt} = n_r(\omega/c)2\pi R$, we have

$$\frac{d\phi_{rt}}{d\omega} = \left(\frac{n_r}{c} + \frac{dn_r}{d\omega}\frac{\omega}{c}\right)2\pi R = \left(n_r + \omega\frac{dn_r}{d\omega}\right)\frac{2\pi R}{c} = \frac{2\pi R}{c/n_g} = T_{rt}.$$

Using the above result, we get $\Delta\phi_{BW} = \Delta\omega_{BW}T_{rt}$.

[†] Strictly speaking, the definition of Q as λ_m divided by the FWHM bandwidth is valid only for classical oscillators of the Lorentzian type (see Equation 2.11). The expression in Equation 2.39 provides a good estimate of the Q factor for low loss and weakly coupled microring resonators ($a_{rt} \approx 1$ and $\tau_1 \approx \tau_2 \approx 1$).

coupling coefficients are assumed to be equal ($\tau_1 = \tau_2 = \tau$) and small ($\tau \approx 1$), we can approximate the Q factor as

$$Q \approx \frac{2\pi^2 n_g R}{\lambda_m} \frac{\tau}{1-\tau^2} \approx \frac{2\pi^2 n_g R}{\lambda_m \kappa^2}. \tag{2.40}$$

The above expression shows that the loaded Q factor increases with the microring radius R and is inversely proportional to the power coupling coefficient κ^2 and the resonant wavelength λ_m.

We can also determine the finesse of the ADMR by dividing the FSR of the microring in Equation 2.5 by the 3 dB or FWHM bandwidth,

$$\mathcal{F} = \frac{\Delta\lambda_{FSR}}{\Delta\lambda_{BW}} = \frac{\pi}{2}\sqrt{F} = \frac{\pi\sqrt{\tau_1 \tau_2 a_{rt}}}{1-\tau_1 \tau_2 a_{rt}}. \tag{2.41}$$

For a symmetrically coupled ADMR with low loss, the above expression simplifies to $\mathcal{F} \approx \pi/\kappa^2$. Since $FE \approx 1/\kappa$, we find that the finesse is related to the field enhancement in the microring by $F = \pi(FE)^2$. From Equations 2.39 and 2.41 we also obtain the relationship between the finesse and the Q factor,

$$\mathcal{F} = \frac{\lambda_m Q}{2\pi n_g R} \approx \frac{Q}{m}, \tag{2.42}$$

where we have used the expression in Equation 2.2 for the resonant wavelength λ_m and assumed that $n_r \approx n_g$.

The physical meaning of the finesse can be deduced by noting that the cavity lifetime of the microring is inversely proportional to the bandwidth $\Delta\omega_{BW}$ (e.g., see Equation. 2.11):

$$\tau_c = \frac{1}{\Delta\omega_{BW}} = \frac{T_{rt}\sqrt{F}}{4} = \frac{T_{rt}\mathcal{F}}{2\pi}. \tag{2.43}$$

The above result shows that the finesse (divided by 2π) gives the number of round-trips a light wave makes around the microring before the stored energy decays to $1/e$ of the initial value due to loss and external coupling. In other words, the finesse gives the cavity lifetime in terms of the number of round-trips. Since there are m periods of wave oscillations per round-trip for resonant mode m, we see from Equation 2.42 that the Q factor gives the cavity lifetime in terms of the number of periods of oscillations.

2.2.2 The all-pass microring resonator

An all-pass microring resonator (APMR) consists of a microring coupled to a single bus waveguide with field coupling coefficient κ, as shown in Figure 2.3a. The structure can be regarded as a special case of the add-drop configuration with $\tau_2 = 1$. Thus, by setting $\tau_1 = \tau$ and $\tau_2 = 1$ in Equation 2.24, we obtain the transfer function of the APMR at the output (or through) port as follows:

$$H_{ap} \equiv \frac{S_t}{S_i} = \frac{\tau - a_{rt}e^{-j\phi_{rt}}}{1 - \tau a_{rt}e^{-j\phi_{rt}}} = \frac{\tau - a_{rt}z^{-1}}{1 - \tau a_{rt}z^{-1}}. \tag{2.44}$$

The power spectral response is given by

$$T_{ap}(\phi_{rt}) = \frac{\tau^2 + a_{rt}^2 - 2\tau a_{rt}\cos\phi_{rt}}{1 + \tau^2 a_{rt}^2 - 2\tau a_{rt}\cos\phi_{rt}} = \frac{T_{ap,min} + F\sin^2(\phi_{rt}/2)}{1 + F\sin^2(\phi_{rt}/2)}, \tag{2.45}$$

$$F = \frac{4\tau a_{rt}}{(1 - \tau a_{rt})^2}, \tag{2.46}$$

$$T_{ap,min} = \left(\frac{\tau - a_{rt}}{1 - \tau a_{rt}}\right)^2. \tag{2.47}$$

The spectral characteristics of an APMR are similar to those of an FP resonator with a perfectly reflecting mirror and a partially reflecting mirror with reflectivity τ, as shown in Figure 2.3b. In the absence of loss ($a_{rt} = 1$), $T_{ap,min} = 1$ and Equation 2.45 gives $T_{ap} = 1$,

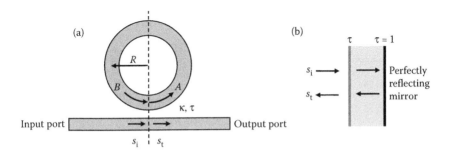

Figure 2.3 (a) Schematic of an APMR with field coupling coefficient κ. (b) In terms of transfer characteristics, the APMR is equivalent to an FP resonator with a perfectly reflecting mirror and a partially reflecting mirror with reflectivity τ.

indicating that power is transmitted through the device without attenuation at all frequencies. The microring thus behaves as an all-pass filter. In the presence of loss, the spectral response is characterized by a transmission dip at the resonant wavelengths. In particular, when $\tau = a_{rt}$, Equation 2.47 gives $T_{ap,min} = 0$, implying that total extinction of the output signal is achieved at the resonant wavelengths. The microring resonator is said to be critically coupled (Yariv 2000). Physically, the input light is destructively interfered by the light coupled out from the microring, resulting in zero power being transmitted at the output port. All the input power in this case is dissipated through the various loss mechanisms in the microring. For other values of τ and a_{rt}, the input power is only partially dissipated and the remaining power is transmitted to the output port. For this more general case, the microring is said to be undercoupled if $\tau > a_{rt}$ and overcoupled if $\tau < a_{rt}$. Typical power spectral responses of an APMR in the regimes of undercoupling, overcoupling, and critical coupling are shown in Figure 2.4a. The associated power enhancements in the microring, $|FE|^2 = |A/s_i|^2$, where A is the field in the microring given by Equation 2.21, are also plotted in Figure 2.4b. We observe that among the different coupling regimes, the critically-coupled microring has the deepest extinction and the largest power enhancement at resonance. We also note that overcoupled APMRs have broader bandwidths than undercoupled microrings because of stronger couplings to the bus waveguide.

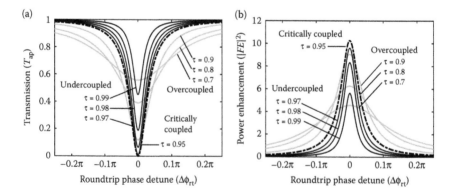

Figure 2.4 (a) Power spectral responses of an APMR for various coupling values and fixed round-trip attenuation $a_{rt} = 0.95$. (b) Power enhancement $|FE|^2$ in the microring as a function of the round-trip phase detune.

The FWHM bandwidth[*] of the resonance spectrum of an APMR can be obtained by setting $\tau_1 = \tau$ and $\tau_2 = 1$ in Equation 2.38,

$$\Delta\lambda_{BW} = \frac{\lambda_m^2}{2\pi^2 n_g R}\frac{1-\tau a_{rt}}{\sqrt{\tau a_{rt}}}. \tag{2.48}$$

For low-loss ($a_{rt} \approx 1$) and weakly coupled ($\tau \approx 1$) microrings, we can approximate the above expression as follows:

$$\Delta\lambda_{BW} \approx \frac{\lambda_m^2}{2\pi^2 n_g R}\frac{1-\tau}{\sqrt{\tau}} \approx \frac{\lambda_m^2}{2\pi^2 n_g R}\frac{1-\tau^2}{\sqrt{\tau}(1+\tau)} \approx \frac{\lambda_m^2 \kappa^2}{4\pi^2 n_g R}. \tag{2.49}$$

The loaded Q factor and the finesse of the APMR are given by

$$Q = \frac{2\pi^2 n_g R}{\lambda_m}\frac{\sqrt{\tau a_{rt}}}{1-\tau a_{rt}} \approx \frac{4\pi^2 n_g R}{\lambda_m \kappa^2}, \tag{2.50}$$

$$\mathcal{F} = \frac{\pi\sqrt{\tau a_{rt}}}{1-\tau a_{rt}} \approx \frac{2\pi}{\kappa^2}, \tag{2.51}$$

where the approximations again apply to low-loss and weakly coupled microrings. Comparing Equations 2.40 and 2.50, we find that the Q factor (and also the finesse) of an APMR is twice that of an ADMR with the same coupling coefficient. This result is expected since an APMR has half the loss due to external coupling as an ADMR.

2.2.3   Phase and GD responses of a microring resonator

The phase response and GD response of an APMR or ADMR are determined by the poles and zeros of the transfer functions of the respective device. Consider first the transfer function at the drop port of an ADMR (Equation 2.23), which can be written as

$$H_d(\phi_{rt}) = -\frac{\kappa_1 \kappa_2 \sqrt{a_{rt}}\, e^{-j\phi_{rt}/2}}{1-p_0^{-1}e^{-j\phi_{rt}}}, \tag{2.52}$$

---

[*] For a resonance dip, the FWHM bandwidth is the bandwidth measured halfway between the maximum transmission (off resonance) and the transmission dip (at the resonant frequency).

where $p_0 = 1/\tau_1\tau_2 a_{rt}$ is the pole in the inverse z-plane ($z^{-1} = e^{-j\phi_{rt}}$). The phase response of H_d is given by

$$\psi_d(\phi_{rt}) = \pi - \frac{\phi_{rt}}{2} - \tan^{-1}\left(\frac{\sin\phi_{rt}}{p_0 - \cos\phi_{rt}}\right), \tag{2.53}$$

where the last term is the phase response due to the pole. The GD response is obtained by differentiating the phase ψ of the transfer function with respect to the angular frequency, $\tau_g = -d\psi/d\omega$. Since $d\phi_{rt}/d\omega = T_{rt}$, we have

$$\tau_g = -\frac{d\psi}{d\omega} = -\frac{d\phi_{rt}}{d\omega}\frac{d\psi}{d\phi_{rt}} = -T_{rt}\frac{d\psi}{d\phi_{rt}}. \tag{2.54}$$

The above expression shows that a pulse passing through the microring device experiences a GD that is enhanced by a factor $S = -d\psi/d\phi_{rt}$ over the round-trip delay T_{rt} of the microring waveguide. The parameter S is called the GD enhancement factor or the slowness factor. By differentiating the phase ψ_d in Equation 2.53 with respect to ϕ_{rt} and using the result in Equation 2.54, we obtain the GD response at the drop port of the ADMR as follows:

$$\tau_{g,d}(\phi_{rt}) = \frac{(p_0^2 - 1)T_{rt}/2}{1 + p_0^2 - 2p_0\cos\phi_{rt}}. \tag{2.55}$$

Writing $\cos\phi_{rt} = 1 - 2\sin^2(\phi_{rt}/2)$, we can also express the above equation in the form

$$\tau_{g,d}(\phi_{rt}) = \frac{S_{p,max}}{1 + F\sin^2(\phi_{rt}/2)}T_{rt}, \tag{2.56}$$

where $F = 4p_0/(p_0 - 1)^2 = 4\tau_1\tau_2 a_{rt}/(1 - \tau_1\tau_2 a_{rt})^2$ is the contrast parameter defined in Equation 2.33, and

$$S_{p,max} = \frac{1}{2}\left(\frac{p_0 + 1}{p_0 - 1}\right) = \frac{1}{2}\left(\frac{1 + \tau_1\tau_2 a_{rt}}{1 - \tau_1\tau_2 a_{rt}}\right) \tag{2.57}$$

is the maximum GD enhancement factor. Equation 2.56 shows that the GD response has the same shape as the power spectral response at the drop port (Equation 2.31). Note that the GD at the drop port is always positive. Physically, the parameter $S_{p,max}$ gives the number of round-trips a pulse makes around the microring at resonance before exiting at the drop port. Like the finesse, it provides a measure of the cavity lifetime in terms of the number of round-trips.[*]

For the through port of the ADMR, we write the transfer function in Equation 2.24 as

$$H_t(\phi_{rt}) = \tau_1 \frac{1 - (\tau_2 a_{rt}/\tau_1)e^{-j\phi_{rt}}}{1 - \tau_1\tau_2 a_{rt}e^{-j\phi_{rt}}} = \tau_1 \frac{1 - z_0^{-1}e^{-j\phi_{rt}}}{1 - p_0^{-1}e^{-j\phi_{rt}}}, \quad (2.58)$$

where $z_0 = \tau_1/\tau_2 a_{rt}$ is the zero and $p_0 = 1/\tau_1\tau_2 a_{rt}$ is the same pole appearing in the drop port transfer function. The phase response at the through port consists of contributions from both the zero and the pole,

$$\psi_t(\phi_{rt}) = \tan^{-1}\left(\frac{\sin\phi_{rt}}{z_0 - \cos\phi_{rt}}\right) - \tan^{-1}\left(\frac{\sin\phi_{rt}}{p_0 - \cos\phi_{rt}}\right). \quad (2.59)$$

To obtain the GD response, we write Equation 2.59 as

$$\psi_t(\phi_{rt}) = \left[\frac{\phi_{rt}}{2} + \tan^{-1}\left(\frac{\sin\phi_{rt}}{z_0 - \cos\phi_{rt}}\right)\right] - \left[\frac{\phi_{rt}}{2} + \tan^{-1}\left(\frac{\sin\phi_{rt}}{p_0 - \cos\phi_{rt}}\right)\right]. \quad (2.60)$$

Noting the resemblance of the terms in the square brackets to the phase response in Equation 2.53, we can write the GD response at the through port as

$$\tau_{g,t}(\phi_{rt}) = -\frac{(z_0^2 - 1)T_{rt}/2}{1 + z_0^2 - 2z_0\cos\phi_{rt}} + \frac{(p_0^2 - 1)T_{rt}/2}{1 + p_0^2 - 2p_0\cos\phi_{rt}}. \quad (2.61)$$

The first term on the right-hand side gives the delay due to the zero, while the second term is the delay due to the pole. Equation 2.61 can also be written as

$$\tau_{g,t}(\phi_{rt}) = -\frac{S_{z,max}}{1 + G\sin^2(\phi_{rt}/2)}T_{rt} + \frac{S_{p,max}}{1 + F\sin^2(\phi_{rt}/2)}T_{rt} = -\tau_{g,zero} + \tau_{g,pole}, \quad (2.62)$$

[*] For high-Q microring resonators ($a_{rt} \approx 1$, $\tau_1 \approx \tau_2 \approx 1$), Equation 2.57 gives $S_{p,max} \approx (1 + \tau^2)/2\kappa^2 \approx 1/\kappa^2$. Since $\mathcal{F} \approx \pi/\kappa^2$, we have $S_{p,max} \approx \mathcal{F}/\pi$.

$$S_{z,max} = \frac{1}{2}\left(\frac{z_0+1}{z_0-1}\right) = \frac{1}{2}\left(\frac{\tau_1+\tau_2 a_{rt}}{\tau_1-\tau_2 a_{rt}}\right), \tag{2.63}$$

$$G = \frac{4z_0}{(z_0-1)^2} = \frac{F}{T_{t,min}}, \tag{2.64}$$

where $T_{t,min}$ is the minimum power transmission at the through port given by Equation 2.30. Equation 2.62 shows that the effect of the pole is to add a positive delay, whereas the effect of the zero is to introduce a negative delay if $\tau_1 > \tau_2 a_{rt}$. At the resonant frequencies, the total GD is positive if $\tau_1 < \tau_2 a_{rt}$ (overcoupling) and negative if $\tau_1 > \tau_2 a_{rt}$ (undercoupling). When $\tau_1 = \tau_2 a_{rt}$ (critical coupling), the GD due to the zero, $\tau_{g,zero}$, approaches negative infinity at the resonant frequencies. However, the power transmitted at the through port also becomes vanishingly small in this case.

For an APMR, the phase and GD responses are given by the same expressions as those at the through port of an ADMR (Equations 2.59 and 2.61), with the pole and zero replaced by $p_0 = 1/\tau a_{rt}$ and $z_0 = \tau/a_{rt}$, respectively. Similar to the through port response of an ADMR, the GD of an APMR is infinite at the resonant frequencies for the critical coupling case ($\tau = a_{rt}$), positive for the undercoupling case ($\tau > a_{rt}$), and negative for the overcoupling case ($\tau < a_{rt}$).

Figure 2.5a and b show the phase and GD responses, respectively, at the drop port and through port of a symmetrically coupled ADMR ($\tau_1 = \tau_2$). In the presence of loss, the device is undercoupled since $\tau_1 > \tau_2 a_{rt}$. We observe in Figure 2.5a that the phases at both

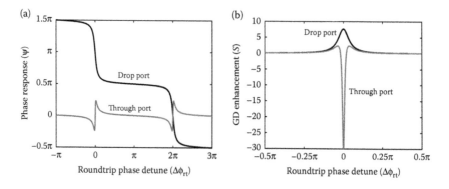

Figure 2.5 (a) Phase response and (b) group delay response ($S = \tau_g/T_{rt}$) at the drop port and through port of an ADMR with coupling coefficients $\kappa_1^2 = \kappa_2^2 = 0.2$ and 5% round-trip power loss ($a_{rt} = 0.975$).

the drop port and through port undergo sharp transitions at the resonances. This rapid phase change near a resonant frequency can be exploited to enhance the performance of phase-sensitive or interferometric devices such as switches, modulators, and sensors. Note that the slope of the phase response at resonance is negative at the drop port and positive at the through port. As a result, we obtain a positive-peak GD response at the former and a negative-peak GD response at the latter, as seen in Figure 2.5b. A negative GD does not mean that an input pulse to the ADMR will arrive at the through port before it enters the input port, implying that causality is violated. Due to the strong GD dispersion (GDD) and attenuation near a resonant frequency, the transmitted pulse at the through port is so much distorted that its arrival time can no longer be unambiguously determined, for example, by tracking the pulse peak. The interpretation of the GD as the transport time of the pulse energy is no longer accurate and, thus, a negative GD value should not be construed as a violation of causality (Brillouin 1960).

The GD at the drop port reaches a positive peak value given by $S_{p,max}T_{rt}$. From Equation 2.57 we find that this value increases for microrings with low loss and weak coupling, which is expected since light in a high-Q microring is trapped for a longer period of time before coupling out onto the bus waveguides. Since the GD response at the drop port has the same shape as the power spectral response, the GD bandwidth is the same as the transmission bandwidth. There is, however, a tradeoff between this bandwidth and the peak GD value. The GD–bandwidth product, $GD \times BW$, which is approximately the area underneath the GD response curve over one FSR, is given by

$$GD \times BW = \int_{FSR} \tau_{g,d} d\omega = -\int_{FSR} \frac{d\psi}{d\omega} d\omega = -\int_{FSR} d\psi. \tag{2.65}$$

Since the phase at the drop port changes by at most $-\pi$ over one FSR (see Figure 2.5a), the above equation implies that the GD–bandwidth product of an ADMR is constrained by π. Thus any increase in the peak GD must be offset by a corresponding decrease in the bandwidth. The GD–bandwidth trade-off is an inherent limitation in all optical delay elements based on passive optical resonators.

In Figure 2.6a and b we plot the phase and GD of an APMR for different coupling values, showing the device responses in the undercoupling, overcoupling, and critical coupling regimes. In all three

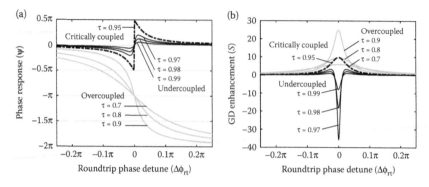

Figure 2.6   (a) Phase response and (b) group delay response ($S = \tau_g/T_{rt}$) of an APMR with various coupling coefficients and fixed round-trip attenuation $a_{rt} = 0.95$.

cases the phase of the output signal undergoes a sharp transition at the resonances. However, in the overcoupling regime, the phase transition slope is positive and becomes steeper for smaller couplings (larger τ values). In the undercoupling regime, the phase has the opposite behavior, exhibiting a positive slope which decreases for increasing τ values. As a result, the GD of an APMR has a positive-peak response for the overcoupling case and a negative-peak response for the undercoupling case. At critical coupling, the phase undergoes a discontinuous jump of π at resonance, which gives an infinite GD value. Note that, in the overcoupling regimes, the phase of the output signal changes by 2π over one FSR, so the GD–bandwidth product of an APMR is constrained by 2π according to Equation 2.65.

The large GD enhancement in a microring resonator makes it useful for realizing optical delay elements or buffers. However, as mentioned above, the GD–bandwidth product places a limit on the achievable peak GD for a given bandwidth. To further increase the GD, we must increase the number of resonators in the device. The analysis and design of arrays of microrings will be addressed in Chapter 3. Another issue with microring delay elements is the GDD, which must be kept small over the device bandwidth to minimize signal distortion. The GDD is defined as

$$D_g = \frac{d\tau_g}{d\omega} \approx -T_{rt}^2 \frac{d^2\psi}{d\phi_{rt}^2},\tag{2.66}$$

where the approximation applies if we neglect GDD of the microring waveguide (i.e., the frequency dependence of T_{rt}). Figure 2.7

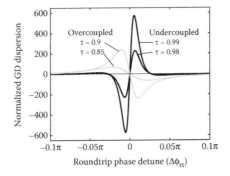

Figure 2.7   Normalized GD dispersion, D_g / T_{rt}^2 of an APMR in the under-coupling and overcoupling regimes. The round-trip attenuation of the microring is fixed at $a_{rt} = 0.95$.

shows the GDD of an APMR in the undercoupling and overcoupling regime. In both cases the GDD is zero at resonance but can rise to a large value over the device bandwidth. Note that the GDD curve has a nearly linear slope at resonance, and the slope is positive for under-coupled microrings and negative for overcoupled devices. Thus, by tuning the coupling coefficient of an APMR, we can vary the GDD slope over a wide range of values in both positive and negative directions. This behavior has been exploited to realized dispersion com-pensators based on microring resonators (Lenz and Madsen 1999).

2.3   Energy Coupling Description of a Microring Resonator

In the previous section, we derived the transfer functions of micro-ring resonators in the add-drop and all-pass configurations using the power coupling approach. The power coupling formalism is rig-orous and general in that no assumption is made about the coupling strengths, as long as the coupling junctions can be modeled by lumped coupling elements.[*] In particular, the formalism can be used to analyze strongly coupled microring resonators, or equivalently, broadband microring devices. Power coupling analysis shows that

---

[*] The power coupling formalism is also valid for microring resonator devices with complex coupling coefficients. For very strong coupling, the transmission coefficient (τ) of the coupling junction will in general be complex, as can be seen from Equation 1.128. Complex coupling coefficients also arise in certain micror-ing coupling topologies, an example of which will be discussed in Section 3.7.

the resonance spectrum of a microring resonator is periodic and has the shape of an Airy function.

An alternative description of microring resonators that is valid for weakly coupled or narrowband devices is the energy coupling formalism, also known as "coupling of modes in time" formalism (Haus 1984, Little et al. 1997). Under the assumption of weak coupling, the field amplitude distribution around the microring can be considered to be uniform. Neglecting the spatial dependence of the field circulating around the microring, we can describe the behavior of the device from the point of view of energy exchange between the resonator and the external waveguides. Specifically, couplings between the microring and bus waveguides are described by energy coupling coefficients, which give the rates of energy coupled into and out of the resonator. In effect, the microring is treated as a classical Lorentz oscillator and, indeed, one can construct an equivalent LC circuit of a microring resonator whose resonant frequency is equal to $\omega_0 = \sqrt{LC}$ (Van 2006). By neglecting the spatial distribution of the field in the microring, the energy coupling formalism provides a simpler model for analyzing microring devices than the power coupling approach.

We consider first an uncoupled microring resonator with resonant frequency ω_0. In the absence of any external coupling, the only source of loss in the resonator is the propagation loss of the microring waveguide. Using Equation 2.8, we can define the decay rate of the mode amplitude due to intrinsic cavity loss as $\gamma_0 = v_g \alpha/2,$[*] where α and v_g are the propagation loss and group velocity, respectively, of the microring waveguide. According to Equation 2.7, we can express the resonant mode in the uncoupled resonator as

$$a(t) = a_0 e^{j\omega_0 t} e^{-\gamma_0 t}, \tag{2.67}$$

where a_0 is the mode amplitude normalized so that $|a_0|^2$ gives the initial stored energy in the microring. By differentiating Equation 2.67 with respect to time, we obtain the equation governing the time decay of mode $a(t)$:

$$\frac{da}{dt} = (j\omega_0 - \gamma_0)a. \tag{2.68}$$

---

[*] To account for the dispersion in the effective index of the microring waveguide, we have replaced the effective index n_r in Equation 2.8 by the group index n_g.

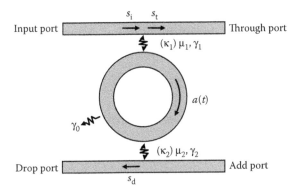

Figure 2.8 Schematic of an ADMR with intrinsic decay rate γ_0, input energy coupling coefficient μ_1 (field coupling κ_1), and output energy coupling coefficient μ_2 (field coupling κ_2).

Consider now a microring resonator coupled to an input waveguide and an output waveguide with input and output field coupling coefficients κ_1 and κ_2, respectively, as shown in Figure 2.8. The external couplings lead to additional losses in the microring due to energy transfer to the two waveguides. Denoting the amplitude decay rates due to coupling to the input and output bus waveguides as γ_1 and γ_2, respectively, we modify Equation 2.68 to read

$$\frac{da}{dt} = (j\omega_0 - \gamma_0 - \gamma_1 - \gamma_2)a. \qquad (2.69)$$

In addition, suppose the input waveguide carries a signal s_i toward the resonator, with the amplitude normalized such that $|s_i|^2$ represents the power flowing in the input waveguide. The input signal supplies energy to the resonator at a rate given by $-j\mu_1 s_i$, where μ_1 is the input energy coupling coefficient and the phase $-j$ is chosen for consistency with the coupling of modes in space formalism (i.e., the term $-j\kappa_1 s_i$ in Equation 2.15). Adding this term to Equation 2.69, we obtain the equation governing the mode amplitude $a(t)$ in an ADMR:

$$\frac{da}{dt} = (j\omega_0 - \gamma_0 - \gamma_1 - \gamma_2)a - j\mu_1 s_i. \qquad (2.70)$$

We can determine the relationship between the energy coupling coefficient μ_1 and the decay rate γ_1 from power conservation consideration (Little et al. 1997). Specifically, we consider a microring resonator with no intrinsic loss ($\gamma_0 = 0$) which is coupled to only an input waveguide ($\gamma_2 = 0$). Suppose the microring is charged to an initial energy $|a_0|^2$ by the input signal s_i. After the input signal is removed ($s_i = 0$), the subsequent energy decay is given by

$$|a(t)|^2 = |a_0|^2 e^{-2\gamma_1 t}. \tag{2.71}$$

The rate of energy decay from the microring is thus

$$\frac{d|a(t)|^2}{dt} = -2\gamma_1 |a(t)|^2. \tag{2.72}$$

The energy leaving the microring is coupled to the input waveguide and is transmitted to the through port, so we must have $|s_t|^2 = -d|a(t)|^2/dt$. Since the power transmitted at the through port is also equal to $|s_t|^2 = \mu_1^2 |a(t)|^2$, we obtain the relation

$$\mu_1^2 = 2\gamma_1. \tag{2.73}$$

Similarly, at the output coupling junction of an ADMR, the energy coupling coefficient μ_2 is related to the output decay rate γ_2 by $\mu_2^2 = 2\gamma_2$.

It is also useful to derive expressions relating the energy coupling coefficients μ_1 and μ_2 to the field coupling coefficients κ_1 and κ_2. First we note that the energy in the microring is related to the power flow by

$$|a(t)|^2 = |A(t)|^2 T_{rt} = |A(t)|^2 \frac{2\pi R}{v_g}, \tag{2.74}$$

where $A(t)$ is the amplitude of the wave traveling in the microring normalized such that $|A(t)|^2$ gives the power flow through any cross section of the microring waveguide at time t. In the absence of an input signal, the power coupled from the microring to the input waveguide is

$$|s_t|^2 = \kappa_1^2 |A(t)|^2 = \mu_1^2 |a(t)|^2. \tag{2.75}$$

Using the relationship between $|a(t)|^2$ and $|A(t)|^2$ in Equation 2.74, we obtain

$$\kappa_1^2 = \mu_1^2 T_{rt} = \frac{2\pi R}{v_g} \mu_1^2. \tag{2.76}$$

Since $T_{rt} = 1/\Delta f_{FSR}$, we can also write $\kappa_1^2 = \mu_1^2/\Delta f_{FSR}$. Similarly, the output field coupling coefficient is related to the output energy coupling coefficient by $\kappa_2^2 = \mu_2^2 T_{rt} = \mu_2^2/\Delta f_{FSR}$.

We now solve Equation 2.70 for the case when the microring is excited by a monochromatic wave of frequency ω, $s_i \sim e^{j\omega t}$. Since the field in the microring will also have the same harmonic time dependence, $a \sim e^{j\omega t}$, we obtain from Equation 2.70

$$j\omega a = (j\omega_0 - \gamma)a - j\mu_1 s_i, \tag{2.77}$$

where $\gamma = \gamma_0 + \gamma_1 + \gamma_2$ is the total decay rate. Solving the above equation for a, we get

$$a = \frac{-j\mu_1 s_i}{j(\omega - \omega_0) + \gamma}, \tag{2.78}$$

which gives the frequency response of the energy amplitude in the microring. Since $s_d = -j\mu_2 a$ and $s_t = s_i - j\mu_1 a$, we obtain the transfer functions at the drop port and through port of the ADMR as follows:

$$\frac{s_d}{s_i} = \frac{-\mu_1\mu_2}{j(\omega - \omega_0) + \gamma}, \tag{2.79}$$

$$\frac{s_t}{s_i} = \frac{j(\omega - \omega_0) + \gamma - \mu_1^2}{j(\omega - \omega_0) + \gamma}. \tag{2.80}$$

Defining the complex frequency variable, $s = j(\omega - \omega_0)$, we can write the above expressions as

$$H_d(s) = \frac{-\mu_1\mu_2}{s + \gamma} = \frac{-\mu_1\mu_2}{s + \gamma_0 + \mu_1^2/2 + \mu_2^2/2}, \tag{2.81}$$

$$H_t(s) = \frac{s + \gamma - \mu_1^2}{s + \gamma} = \frac{s + \gamma_0 - \mu_1^2/2 + \mu_2^2/2}{s + \gamma_0 + \mu_1^2/2 + \mu_2^2/2}. \tag{2.82}$$

The above expressions have the form of a transfer function of an analog filter* with a single pole at $s = -\gamma$. We also note that the drop

---

* The inverse Laplace transforms of $H_d(s)$ and $H_t(s)$ give the impulse responses at the drop port and through port, respectively, of the ADMR.

port transfer function has no zero, whereas the through port transfer function has a zero at $s = \mu_1^2 - \gamma$.

The power spectral responses of the ADMR are obtained by taking the absolute square of Equations 2.79 and 2.80:

$$T_d(\omega) = \left|\frac{s_d}{s_i}\right|^2 = \frac{\mu_1^2 \mu_2^2}{(\omega - \omega_0)^2 + \gamma^2}, \tag{2.83}$$

$$T_t(\omega) = \left|\frac{s_t}{s_i}\right|^2 = \frac{(\omega - \omega_0)^2 + (\gamma - \mu_1^2)^2}{(\omega - \omega_0)^2 + \gamma^2}. \tag{2.84}$$

Equation 2.83 indicates that the drop port response has the shape of a Lorentzian resonance centered at the resonant frequency ω_0. The FWHM bandwidth is given by $\Delta\omega_{BW} = 2\gamma$, or in terms of wavelength:

$$\Delta\lambda_{BW} = \frac{\lambda_0^2 \gamma}{\pi c}. \tag{2.85}$$

To determine the Q factor of the ADMR, we can use the expression $Q = \lambda_0/\Delta\lambda_{BW}$ to get $Q = \pi c/\lambda_0 \gamma = \omega_0/2\gamma$. The same formula can also be obtained directly from the definition of the Q factor:

$$Q = \omega_0 \frac{\text{Stored energy}}{\text{Power loss}}. \tag{2.86}$$

Since the stored energy in the microring is $|a|^2$ and the total power loss is $2\gamma|a|^2$, we have

$$Q = \omega_0 \frac{|a|^2}{2\gamma|a|^2} = \frac{\omega_0}{2\gamma}. \tag{2.87}$$

If we define the Q factor due to intrinsic loss as $Q_0 = \omega_0/2\gamma_0$, and the Q factor due to coupling to each of the bus waveguides as $Q_1 = \omega_0/2\gamma_1 = \omega_0/\mu_1^2$ and $Q_2 = \omega_0/2\gamma_2 = \omega_0/\mu_2^2$; then, from Equation 2.87, we have

$$\frac{1}{Q} = \frac{1}{Q_0} + \frac{1}{Q_1} + \frac{1}{Q_2}. \tag{2.88}$$

For an ADMR with low loss ($\gamma_0 \approx 0$) and symmetric coupling ($\mu_1 = \mu_2 = \mu$), we have $Q_0 \to \infty$, $Q_1 = Q_2 = \omega_0/\mu^2$, so

$$Q = \frac{\omega_0}{2\mu^2} = \frac{2\pi^2 n_g R}{\lambda_0 \kappa^2},$$
(2.89)

which agrees with the expression for the loaded Q factor in Equation 2.40.

For an APMR with energy coupling coefficient μ, the transfer function can be obtained by setting $\mu_1 = \mu$ and $\mu_2 = 0$ in Equation 2.82

$$H_{ap}(s) = \frac{s+\gamma-\mu^2}{s+\gamma} = \frac{s+\gamma_0-\mu^2/2}{s+\gamma_0+\mu^2/2}.$$
(2.90)

From the above equation we see that complete extinction of the transmitted signal is achieved at resonance if the decay rate due to external coupling, $\gamma_e = \mu^2/2$, is equal to the decay rate due to intrinsic loss, $\gamma_e = \gamma_0$. This is the critical coupling condition. The APMR is undercoupled if $\gamma_e < \gamma_0$ and overcoupled if $\gamma_e > \gamma_0$.

We can also obtain the phase and GD responses of an ADMR from its transfer functions in Equations 2.81 and 2.82. The phase responses at the drop port and through port are given by

$$\psi_d(\omega) = \pi - \tan^{-1}\left(\frac{\omega-\omega_0}{\gamma}\right),$$
(2.91)

$$\psi_t(\omega) = \tan^{-1}\left(\frac{\omega-\omega_0}{\gamma-\mu_1^2}\right) - \tan^{-1}\left(\frac{\omega-\omega_0}{\gamma}\right).$$
(2.92)

The GD responses are obtained by differentiating the phase with respect to ω. The results for the GD at the drop port and through port are

$$\tau_{g,d}(\omega) = \frac{\gamma}{(\omega-\omega_0)^2+\gamma^2},$$
(2.93)

$$\tau_{g,t}(\omega) = -\frac{\gamma-\mu_1^2}{(\omega-\omega_0)^2+(\gamma-\mu_1^2)^2} + \frac{\gamma}{(\omega-\omega_0)^2+\gamma^2}.$$
(2.94)

From Equation 2.93 we find that the peak GD at the drop port is $\tau_{g,max} = 1/\gamma$, which is equal to the decay time of the wave amplitude in the microring. The FWHM bandwidth of the GD response at the drop port is $\Delta\omega_{BW} = 2\gamma$, so the GD–bandwidth product is equal to 2. Finally, we note that the phase and GD responses of an APMR are also given by Equations 2.92 and 2.94, respectively, with $\mu_1 = \mu$ and $\mu_2 = 0$.

2.4 Relationship between Energy Coupling and Power Coupling Formalisms

Further insight into the energy coupling formalism can be gained by establishing a direct relationship between the power coupling and energy coupling formalisms. In particular, we will show that in the limit of low loss and weak coupling, the power coupling formalism reduces to the energy coupling formalism. In Section 2.2, we found that the power-normalized field inside an ADMR is given by (Equation 2.21):

$$A = \frac{-j\kappa_1 s_i}{1 - \tau_1 \tau_2 a_{rt} e^{-j\phi_{rt}}}.$$ (2.95)

Writing the round-trip phase in the microring as $\phi_{rt} = 2m\pi + \Delta\phi_{rt}$, where $\Delta\phi_{rt}$ is the phase detune from the resonance, we have

$$e^{-j\phi_{rt}} = e^{-j\Delta\phi_{rt}} = e^{-j\Delta\omega T_{rt}},$$ (2.96)

where $\Delta\omega = \omega - \omega_0$ and $T_{rt} = 2\pi R/v_g$ is the round-trip time of the microring. The round-trip amplitude attenuation factor a_{rt} can likewise be written as

$$a_{rt} = e^{-\alpha\pi R} = e^{-\gamma_0 T_{rt}},$$ (2.97)

where $\gamma_0 = \alpha v_g/2$ is the intrinsic decay rate due to loss in the microring. The denominator of Equation 2.95 can now be expressed as

$$1 - \tau_1 \tau_2 a_{rt} e^{-j\phi_{rt}} = 1 - \exp[\ln(\tau_1) + \ln(\tau_2) - (\gamma_0 + j\Delta\omega)T_{rt}].$$ (2.98)

For weak couplings we have

$$\ln(\tau_1) = \frac{1}{2}\ln(1 - \kappa_1^2) \approx \frac{-\kappa_1^2}{2} = -\frac{\mu_1^2}{2}T_{rt},$$ (2.99)

with a similar expression for $\ln(\tau_2)$. With these approximations, Equation 2.98 becomes

$$1-\tau_1\tau_2 a_{rt}e^{-j\phi_{rt}} \approx \left[\frac{\mu_1^2}{2}+\frac{\mu_2^2}{2}+(\gamma_0+j\Delta\omega)\right]T_{rt}, \tag{2.100}$$

where we have also used the small argument approximation for the exponential function, which is valid for weak couplings, small loss and small frequency detunes around the microring resonance. Substituting Equation 2.100 into Equation 2.95, we get

$$A \approx \frac{-j(\mu_1\sqrt{T_{rt}})s_i}{[\mu_1^2/2+\mu_2^2/2-(\gamma_0+j\Delta\omega)]T_{rt}}. \tag{2.101}$$

Since the energy amplitude a in the microring is related to the power-normalized field A by $a = A\sqrt{T_{rt}}$, we obtain from Equation 2.101

$$a = \frac{-j\mu_1 s_i}{(j\Delta\omega+\gamma_0)+\mu_1^2/2+\mu_2^2/2}. \tag{2.102}$$

The above expression for the energy amplitude in the microring is the same as Equation 2.78 obtained using the energy coupling formalism. We thus conclude that the energy coupling formalism is an approximation of the power coupling formalism under the weak coupling and low-loss (or equivalently, narrowband) condition.

Figure 2.9 compares the spectral responses of a relatively strongly coupled ADMR obtained using the power coupling and

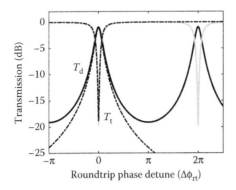

Figure 2.9   Power spectral responses at the drop port (T_d) and through port (T_t) of an ADMR with power coupling coefficients $\kappa_1^2 = \kappa_2^2 = 0.2$ and 5% round-trip power loss ($a_{rt} = 0.975$). Solid lines are spectra obtained using the power coupling formalism; dashed lines are spectra obtained using the energy coupling formalism.

energy coupling approaches. The power coupling coefficients are $\kappa_1^2 = \kappa_2^2 = 0.2$ and the round-trip power loss is 5%. One notable difference between the two results is that the spectral responses obtained from power coupling are periodic, whereas the energy coupling formalism can model only one resonance (the center resonance at $\Delta\phi_{rt} = 0$). In addition, we observe that the discrepancy between the two sets of spectra becomes more pronounced as we move away from the center resonance. Thus the energy coupling approximation is only good for frequencies near a microring resonance.

2.5 Summary

In this chapter, we developed two alternate analytical models for a microring resonator: a rigorous model based on power coupling and an approximate but simpler model based on energy coupling. Using these models, we obtained the transfer functions of the microring in the add-drop and all-pass configurations, and derived useful formulas for determining important quantities characterizing the performance of these devices. In subsequent chapters, we will also make extensive use of both the power coupling and energy coupling formalisms to analyze and design more complex device configurations and more advanced applications of microring resonators. More specifically, we will extend the analysis techniques developed in this chapter to design optical filters based on coupled microring resonators in Chapter 3. The behaviors of microring resonators in the presence of optically-induced and electrically-induced material nonlinearities will be studied in Chapters 4 and 5, respectively.

References

Brillouin, L. 1960. *Wave Propagation and Group Velocity*. New York: Academic Press.

Haus, H. A. 1984. *Waves and Fields in Optoelectronics*. Englewood Cliffs, NJ: Prentice-Hall.

Lenz, G., Madsen, C. K. 1999. General optical all-pass filter structures for dispersion control in WDM systems. *J. Lightwave Technol.* 17(7): 1248–1254.

Little, B. E., Chu, S. T., Haus, H. A., Foresi, J., Laine, J.-P. 1997. Microring resonator channel dropping filters. *J. Lightwave Technol.* 15(6): 998–1005.

Madsen, C. K., Zhao, J. H. 1999. *Optical Filter Design and Analysis: A Signal Processing Approach.* New York: John Wiley & Sons Inc.

Van, V. 2006. Circuit-based method for synthesizing serially coupled microring filters. *J. Lightwave Technol.* 24(7): 2912–2919.

Yariv, A. 2000. Universal relations for coupling of optical power between microresonators and dielectric waveguides. *Electron. Lett.* 36(4): 321–322.

CHAPTER 3

Coupled Microring Optical Filters

One of the most important applications of microring resonators is the synthesis of optical filters. In the broadest sense, a filter is a structure or device that alters the amplitude, phase, or group-delay characteristics of an input signal in a desired fashion. Owing to their compact sizes and strongly dispersive characteristics near a resonant frequency, microring resonators have been shown to be particularly versatile elements for constructing high-order integrated optical filters.

This chapter introduces the theories and techniques for analyzing and designing coupled microring resonator (CMR) optical filters. We will begin in Section 3.1 with a discussion of infinite arrays of microring resonators and show how the periodicity of these structures gives rise to frequency responses with passbands and stopbands. We will then turn our focus to finite arrays of CMRs and study how their spectral characteristics can be tailored to achieve a desired filter response. Representations of microring filters as analog and digital IIR filters are discussed in Section 3.2 along with a review of the general forms of the transfer functions of each type of filters. The remainder of the chapter will be devoted to the analysis and design of specific microring filter configurations. Although numerous variations of microring filter designs have been proposed and studied in the literature, we will restrict our treatment to the more universal types, namely, cascaded all-pass and add-drop microring filters, and serially coupled microring filters and their generalizations to 2D coupling topologies. With the exception of the cascaded add-drop microring array, each of these structures can exactly realize a specific type of transfer function of a given order. For each type of filters, we will develop techniques for analyzing and designing high-order optical filters based on the power coupling formalism. Simpler techniques based on energy coupling will also be developed for coupled microring networks of 1D and 2D coupling topologies.

3.1   Periodic Arrays of Microring Resonators

Periodic microring arrays of infinite length behave as 1D artificial media with distributed resonances. These structures have unique photonic band structures which give rise to many interesting light propagation characteristics. Three particular configurations—the all-pass microring array, the cascaded add-drop microring array, and the serially coupled microring array—have been well studied. The serially coupled microring array is also known as CROW (Coupled Resonator Optical Waveguides) (Yariv et al. 1999), while parallel cascaded arrays of all-pass and add-drop microrings are sometimes called single-channel and double-channel SCISSORs (for side-coupled integrated spaced sequence of resonators) (Heebner et al. 2004), respectively. The double-channel SCISSOR and CROW structures share the common feature that the distributed resonances and feedback give rise to photonic band structures consisting of alternating transmission bands and stopbands. Strong dispersion occurs near the band edges, giving rise to slow light effects which can be exploited for applications in optical delay lines and group velocity dispersion engineering (Melloni et al. 2003).

In general, the dispersion characteristics of a periodic array of distributed resonators can be analyzed using the Bloch matrix formalism (Yeh et al. 1977, Heebner et al. 2004). The array can be modeled by a periodic sequence of unit elements as shown in Figure 3.1, with each element described by a transfer matrix (or transmission matrix) $\mathbf{T}$ defined as

$$\begin{bmatrix} a_{k+1} \\ b_{k+1} \end{bmatrix} = \begin{bmatrix} T_{11} & T_{12} \\ T_{21} & T_{22} \end{bmatrix} \begin{bmatrix} a_k \\ b_k \end{bmatrix}. \tag{3.1}$$

For an infinite array of identical elements with spatial periodicity Λ, Bloch's theorem states that the output fields and input fields of each unit element are related by

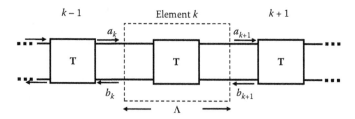

Figure 3.1 Schematic of an infinite periodic array of identical unit elements, each described by transfer matrix $\mathbf{T}$.

$$\begin{bmatrix} a_{k+1} \\ b_{k+1} \end{bmatrix} = \begin{bmatrix} a_k \\ b_k \end{bmatrix} e^{-j\beta\Lambda}, \tag{3.2}$$

where β is the effective propagation constant or Bloch vector. If there is loss in the array, the propagation constant will be complex. Upon substituting Equation 3.2 into Equation 3.1, we obtain

$$\begin{bmatrix} T_{11} & T_{12} \\ T_{21} & T_{22} \end{bmatrix} \begin{bmatrix} a_k \\ b_k \end{bmatrix} = \begin{bmatrix} a_k \\ b_k \end{bmatrix} e^{-j\beta\Lambda}, \tag{3.3}$$

which implies that $\lambda = e^{-j\beta\Lambda}$ are the eigenvalues of the matrix **T**. These eigenvalues are the roots of the characteristic equation

$$\Delta_T e^{j\beta\Lambda} + e^{-j\beta\Lambda} = T_{11} + T_{22}, \tag{3.4}$$

where $\Delta_T = T_{11}T_{22} - T_{12}T_{21}$ is the determinant of the transfer matrix. For a lossless unit element, the matrix **T** is unitary, that is, its determinant is equal to 1. In this case Equation 3.4 can be simplified to read

$$\cos(\beta\Lambda) = \frac{1}{2}(T_{11} + T_{22}). \tag{3.5}$$

In addition, for symmetrical networks we have $T_{22} = T_{11}^*$, in which case Equation 3.5 further reduces to

$$\cos(\beta\Lambda) = \text{Re}\{T_{11}\}. \tag{3.6}$$

Equation 3.6 gives the dispersion relation of a periodic array of unit elements characterized by transfer matrix **T**. Note that the dispersion relation depends only on the parameters T_{11} and T_{22}. In the following subsections, we examine in detail the transmission characteristics of several common types of periodic microring arrays.

3.1.1 Periodic arrays of APMRs

An all-pass microring array consists of an infinite sequence of identical APMRs, as depicted in Figure 3.2. Each microring has radius R and is coupled to a common bus waveguide with field coupling coefficient κ (transmission coefficient τ). The microring waveguide is

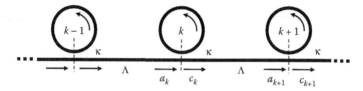

Figure 3.2   Schematic of a periodic array of APMRs.

assumed to have effective index n_r and propagation loss α. The output of each microring is fed forward to the next stage, so there is no feedback in the array. The absence of distributed feedback implies that the structure has no Bragg bandgap. Each unit element in the array consists of an APMR and a bus waveguide of length Λ. The output of element k is given by $a_{k+1} = T_{11}a_k$, where T_{11} is the product of the transfer function of the APMR and a phase delay $e^{-j\theta}$ due to the bus waveguide:

$$T_{11} = H_{ap}e^{-j\theta} = \frac{\tau - a_{rt}e^{-j\phi}}{1 - \tau a_{rt}e^{-j\phi}}e^{-j\theta} \equiv e^{-j\Phi}. \tag{3.7}$$

In the above equation, $\phi = 2\pi n_r kR$ and $a_{rt} = \exp(-\pi\alpha R)$ are the round-trip phase and amplitude attenuation, respectively, of the microring. The angle θ is given by $\theta = n_b k\Lambda$, where n_b is the effective index of the bus waveguide. The phase angle Φ of T_{11} defined in Equation 3.7 is real if there is no loss in the microrings ($|H_{ap}| = 1$); otherwise it will be complex. Applying Bloch's theorem, we have

$$a_{k+1} = a_k e^{-j\Phi} = a_k e^{-j\beta\Lambda}, \tag{3.8}$$

which gives $\beta\Lambda = \Phi$.

For a lossless APMR ($a_{rt} = 1$), the transfer function can be expressed as

$$H_{ap} = \frac{\tau - e^{-j\phi}}{1 - \tau e^{-j\phi}} = -e^{-j\phi}\frac{1 - \tau e^{j\phi}}{1 - \tau e^{-j\phi}}, \tag{3.9}$$

from which we obtain the phase angle of H_{ap} as follows:

$$\psi_{ap} = \pi - \phi - 2\tan^{-1}\left(\frac{\sin\phi}{\tau - \cos\phi}\right). \tag{3.10}$$

The dispersion relation for the Bloch wave vector of the APMR array is thus $\beta\Lambda = \Phi = \theta - \psi_{ap}$, or

$$\beta = \frac{1}{\Lambda}\left[\phi + 2\tan^{-1}\left(\frac{\sin\phi}{\tau - \cos\phi}\right) - \pi + \theta\right]. \tag{3.11}$$

If there is loss in the microrings, we can obtain the phase angle ψ_{ap} from the more general expression in Equation 2.59 with $z_0 = \tau/a_{rt}$ and $p_0 = 1/\tau a_{rt}$. The dispersion relation of the APMR array is given by $\mathrm{Re}\{\beta\Lambda\} = \theta - \psi_{ap}$, or

$$\mathrm{Re}\{\beta\} = \frac{1}{\Lambda}\left[\tan^{-1}\left(\frac{\tau a_{rt}\sin\phi}{1 - \tau a_{rt}\cos\phi}\right) - \tan^{-1}\left(\frac{a_{rt}\sin\phi}{\tau - a_{rt}\cos\phi}\right) + \theta\right]. \tag{3.12}$$

Note that except for an additional phase shift θ due to the bus wave-guide, the phase response per period of the APMR array is the same as that of a single APMR.

Figure 3.3a and b compare the dispersion diagram (plot of ϕ vs. $\beta\Lambda$) of an array of overcoupled microrings with that of an array of undercoupled microrings. In both arrays the length Λ of the bus waveguide is chosen to be equal to the microring circumference so that we have $\theta = \phi$ (assuming that $n_r = n_b$). We will also assume n_r and n_b to have negligible dispersion so that the group index of the waveguide is the same as the effective index. The overcoupled microrings have transmission coefficient $\tau = 0.9$ and round-trip attenuation $a_{rt} = 0.99$, whereas the parameters for the undercoupled microrings are $\tau = 0.9$ and $a_{rt} = 0.85$. Also shown in the figures are the normalized group index, $n_g/n_r = \Lambda(d\beta/d\phi)$, and the power transmission through 10 periods of the array. The transmission is computed from $|T_{11}|^{2N}$, where $N = 10$ is the number of periods. From the plots we observe that the dispersion characteristics of the arrays are dominated by the resonances of the all-pass microrings, with the group index exhibiting strong dispersion near each microring resonance. For the array of overcoupled AMPRs, the group index is positive and becomes significantly larger than the effective index of the microring waveguide near each resonance, indicating that light is slowed down as it propagates through the array. For the array of undercoupled APMRs, the group index becomes negative over a narrow band of frequencies centered around each resonance. This is the regime of fast light, or superluminal propagation, where

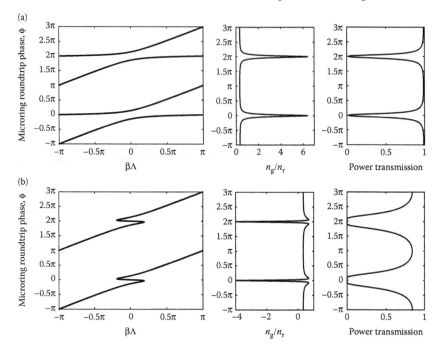

Figure 3.3 Dispersion characteristics of APMR arrays with $\Lambda = 2\pi R$: (a) array of overcoupled microrings ($\tau = 0.9$; $a_{rt} = 0.99$), (b) array of under-coupled microrings ($\tau = 0.9$; $a_{rt} = 0.85$). Left panel: dispersion diagram; center panel: normalized group index n_g/n_r; right panel: power transmission through 10 periods. (After Heebner, J. E., et al., 2004, *J. Opt. Soc. Am. B* 21(10): 1818–1832.)

a light pulse can appear to travel backward or take less time to traverse a section of the array than in a straight waveguide of the same length. However, the transmitted light signal is also severely attenuated and distorted due to the strong group-delay dispersion in the stopband.

Note that the dispersion diagrams of both APMR arrays do not show any bandgap since there is no feedback in the arrays. For an ideal lossless APMR array, the transmission is unity for all frequencies. However, in practical structures, a slight amount of microring loss can cause sharp transmission dips to occur around the resonances, as evidenced in the transmission plot of the overcoupled microring array in Figure 3.3a. As the microring loss increases, the stopband widens further, as seen for the case of the undercoupled array in Figure 3.3b. To access the slow light (or fast light) regime, one would have to operate within the stopband, which results in significant attenuation of the signal. This is a major drawback of APMR arrays for slow light applications. Nevertheless, an array of 56 APMRs has been experimentally demonstrated in the SOI material system for on-chip optical buffer applications (Xia et al. 2007b).

3.1.2 Periodic arrays of ADMRs

Figure 3.4 shows a schematic of a periodic array of parallel cascaded ADMRs. We assume the ADMRs to be identical with equal input and output field coupling coefficients κ (transmission coefficients τ). Adjacent ADMRs are connected via two parallel input and output bus waveguides of length Λ, which is assumed to be greater than the microring diameter so that there is no direct coupling between two adjacent microrings. A wave traveling in the input bus waveguide will be partially coupled to each microring and subsequently transmitted to the output bus waveguide, where it propagates backward as a reflected wave. The array can thus be regarded as a periodic grating in which the microrings act as reflecting elements. Therefore, we expect the structure to have Bragg bandgaps centered at wavelengths satisfying the condition $2\Lambda = m_B(\lambda/n_b)$, where m_B is an integer and n_b is the effective index of the bus waveguides. In addition, since the ADMRs are strongly reflecting at wavelengths corresponding to the microring resonances, there are also stopbands occurring at wavelengths satisfying the resonance condition $2\pi R = m_R(\lambda/n_r)$, where m_R is an integer and n_r is the effective index of the microring waveguide. Overlapping between the Bragg bandgaps and microring stopbands occurs if the ratio $\Lambda/\pi R$ is equal to $n_r m_B/n_b m_R$.

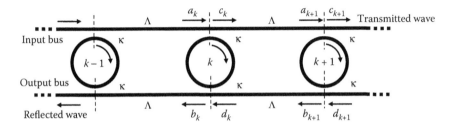

Figure 3.4 Schematic of a periodic array of cascaded add-drop microring resonators.

Each unit element in the array consists of an ADMR and two parallel waveguide segments of length Λ. For simplicity, we assume that the microrings are lossless. Using Equations 2.23 and 2.24 for the drop port and through port transfer functions of an ADMR, we can write, for each microring k,

$$\begin{bmatrix} b_k \\ c_k \end{bmatrix} = \frac{1}{1-\tau^2 e^{-j\phi}} \begin{bmatrix} -\kappa^2 e^{-j\phi/2} & \tau(1-e^{-j\phi}) \\ \tau(1-e^{-j\phi}) & -\kappa^2 e^{-j\phi/2} \end{bmatrix} \begin{bmatrix} a_k \\ d_k \end{bmatrix}, \tag{3.13}$$

where ϕ is the microring round-trip phase. The above matrix equation can be recast in the form of a transfer matrix $\mathbf{T}_{ad}$ of the ADMR as

$$\begin{bmatrix} c_k \\ d_k \end{bmatrix} = \frac{1}{\tau(1-e^{-j\phi})} \begin{bmatrix} (\tau^2 - e^{-j\phi}) & -\kappa^2 e^{-j\phi/2} \\ \kappa^2 e^{-j\phi/2} & (1-\tau^2 e^{-j\phi}) \end{bmatrix} \begin{bmatrix} a_k \\ b_k \end{bmatrix} \equiv \mathbf{T}_{ad} \begin{bmatrix} a_k \\ b_k \end{bmatrix}. \tag{3.14}$$

The transfer matrix of the two parallel bus waveguides of length Λ is given by

$$\begin{bmatrix} a_{k+1} \\ b_{k+1} \end{bmatrix} = \begin{bmatrix} e^{-j\theta} & 0 \\ 0 & e^{j\theta} \end{bmatrix} \begin{bmatrix} c_k \\ d_k \end{bmatrix} \equiv \Lambda \begin{bmatrix} c_k \\ d_k \end{bmatrix}, \tag{3.15}$$

where $\theta = n_b k \Lambda$. From Equations 3.14 and 3.15, we obtain the transfer matrix of each unit element in the array as $\mathbf{T} = \Lambda \mathbf{T}_{ad}$, which is a unitary matrix with the elements T_{11} and T_{22} given by

$$T_{11} = \frac{\tau^2 - e^{-j\phi}}{\tau(1-e^{-j\phi})} e^{-j\theta} = T_{22}^*. \tag{3.16}$$

Substituting the above expression for T_{11} into Equation 3.6, we arrive at the following dispersion relation for the ADMR array:

$$\beta = \pm \frac{1}{\Lambda} \cos^{-1} \left\{ \frac{1}{\sin(\phi/2)} \left[\frac{\tau}{2} \sin\left(\frac{\phi}{2} + \theta \right) + \frac{1}{2\tau} \sin\left(\frac{\phi}{2} - \theta \right) \right] \right\}. \tag{3.17}$$

The plus and minus signs correspond to the forward- and backward-propagating Bloch waves, respectively, in the array.

Figure 3.5 shows the dispersion diagram (plot of ϕ vs. $\beta\Lambda$) of an array of lossless ADMRs with $\tau = 0.8$, $\Lambda/2\pi R = 2/3$, and effective index $n_r = n_b$, which is assumed to be dispersionless. The normalized

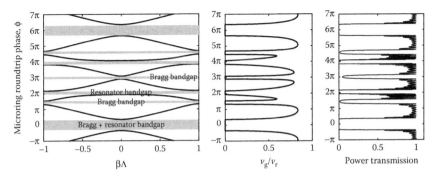

Figure 3.5 Dispersion characteristics of a lossless ADMR array with $\tau = 0.8$ and $\Lambda/2\pi R = 2/3$. Left panel: dispersion diagram; center panel: normalized group velocity (v_g/v_r); right panel: power transmission through a 10-period array. (After Heebner, J. E., et al., 2004, *J. Opt. Soc. Am. B* 21(10): 1818–1832.)

group velocity, v_g/v_r where $v_r = c/n_r$, and the power transmission[*] through 10 periods are also shown. The dispersion diagram displays a complex pattern consisting of stopbands due to both distributed feedback and microring resonances. Note that the Bragg stopbands are of the direct bandgap type since the band maxima align with the band minima, whereas those due to the microring resonances are of the indirect type (Heebner et al. 2004). The two types of stopbands overlap every 6π change in the microring round-trip phase, which corresponds to three free spectral ranges (FSRs) of the resonator. From the plot of the normalized group velocity, we observe that within each transmission band there is a reduction in the group velocity of the Bloch mode compared to the group velocity in the waveguide $(v_g < v_r)$. This slow light effect is caused by light being trapped in the microrings for long periods of time (equal to the microring lifetime) and/or undergoing multiple Bragg reflections as it propagates along the array. The group velocity can be further reduced by decreasing the microring coupling strength κ to increase the lifetime of each resonator.

The plot of the power transmission through the array shows that within each passband, the transmission reaches unity but also exhibits large ripples. These ripples are caused by the fact that the array in the example is truncated with a finite number of periods

---

[*] Formulas for computing the transmission and reflection spectra of an ADMR array of finite length are derived in Section 3.5.

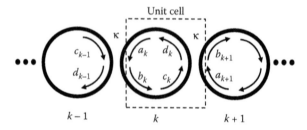

Figure 3.6   Schematic of a CROW lattice consisting of a periodic array of serially coupled microring resonators.

$(N = 10)$, resulting in reflections of the Bloch waves at the terminations. The transmission spectra flatten out in the limit N becomes very large. For an ADMR array of finite length, the ripples in the transmission response can be reduced by applying apodization to the coupling coefficients. We will look at the analysis and design of this type of structures in more detail in Section 3.5.

3.1.3   Arrays of serially coupled microring resonators

Figure 3.6 shows a schematic of an infinite chain of serially coupled microring resonators, which is also known as a Coupled Resonator Optical Waveguide (CROW) (Yariv et al. 1999). The microrings are assumed to be identical with radius R and adjacent microrings are directly coupled to each other via the field coupling coefficient κ (transmission coefficient τ). Since energy in a microring can be coupled forward and backward to its two neighbors, the CROW array supports both forward- and backward-propagating Bloch modes. At resonance, energy transport along the array is considerably slowed down since it takes time for light to "charge" each microring before moving to the next. At off-resonance wavelengths, light is prevented from building up in the microrings and hence cannot propagate in the array. Thus, we expect the frequency response of the array to have stopbands separating the microring resonances. Indeed, a CROW array can be regarded as a distributed feedback grating in which each coupling junction acts as a partial reflector. In this case, transmission bands occur when the Bragg condition $2\Lambda = m(\lambda/n_r)$ is satisfied,[*] where $\Lambda = \pi R$ is the length of each period.

---

[*] The Bragg condition $2\Lambda = m\lambda/n_r$ corresponds to passbands rather than stopbands in this case because the transmission coefficient of the partial reflector is imaginary (equal to $-j\kappa$).

Note that this Bragg condition is also the resonance condition of the microrings.

To derive the dispersion relation of an infinite CROW array, we define each unit element in the array as consisting of a microring and a coupling junction, as shown in Figure 3.6 (Poon et al. 2004). For simplicity, we will assume that the microrings have no loss. Within microring k, the fields $[c_k, d_k]$ are related to $[a_k, b_k]$ by

$$
\begin{bmatrix} c_k \\ d_k \end{bmatrix} = \begin{bmatrix} 0 & e^{-j\phi/2} \\ e^{j\phi/2} & 0 \end{bmatrix} \begin{bmatrix} a_k \\ b_k \end{bmatrix} \equiv \Lambda \begin{bmatrix} a_k \\ b_k \end{bmatrix},
\tag{3.18}
$$

where ϕ is the microring round-trip phase. At the coupling junction between microrings k and $k + 1$, we have the relations

$$
d_k = \tau c_k - j\kappa a_{k+1},
\tag{3.19}
$$

$$
b_{k+1} = \tau a_{k+1} - j\kappa c_k.
\tag{3.20}
$$

The above expressions can be rearranged to give the transfer matrix $\mathbf{K}$ of the coupling junction as follows:

$$
\begin{bmatrix} a_{k+1} \\ b_{k+1} \end{bmatrix} = \frac{1}{j\kappa} \begin{bmatrix} \tau & -1 \\ 1 & -\tau \end{bmatrix} \begin{bmatrix} c_k \\ d_k \end{bmatrix} \equiv \mathbf{K} \begin{bmatrix} c_k \\ d_k \end{bmatrix}.
\tag{3.21}
$$

The transfer matrix of each unit element in the CROW array can be obtained using Equations 3.18 and 3.21:

$$
\mathbf{T} = \mathbf{K}\Lambda = \frac{1}{j\kappa} \begin{bmatrix} -e^{j\phi/2} & \tau e^{-j\phi/2} \\ -\tau e^{j\phi/2} & e^{-j\phi/2} \end{bmatrix}.
\tag{3.22}
$$

The above matrix is unitary with $T_{11} = T_{22}^* = -e^{j\phi/2}/j\kappa$. Using Equation 3.6, we obtain the following dispersion relation for the CROW array:

$$
\beta = \pm \frac{1}{\Lambda} \cos^{-1} \left[-\frac{1}{\kappa} \sin(\phi/2) \right].
\tag{3.23}
$$

The plus and minus signs correspond to the forward- and backward-propagating Bloch waves, respectively, in the array.

Figure 3.7 shows the dispersion diagram (plot of ϕ vs. $\beta\Lambda$) of a lossless CROW array with $\tau = 0.8$. We assume the effective index n_r

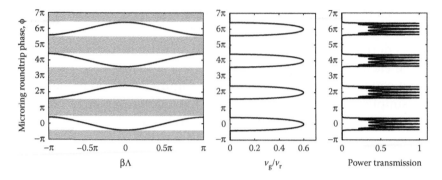

Figure 3.7 Dispersion characteristics of a lossless CROW with $\tau = 0.8$. Left panel: dispersion diagram; center panel: normalized group velocity (v_g/v_r); right panel: power transmission through five microrings. (After Heebner, J. E., et al., 2004, *J. Opt. Soc. Am. B* 21(10): 1818–1832.)

of the microring waveguide to have negligible dispersion so that its group index is also the same as the effective index. Also shown in the figure are the normalized group velocity, v_g/v_r where $v_r = c/n_r$, and the power transmission spectrum[*] through a CROW array consisting of $N = 5$ microrings. The dispersion diagram shows the periodic formation of a passband and a stopband for every 2π change in the microring round-trip phase (or one microring FSR). In the plot of the normalized group velocity, we observe that within each passband the group velocity is reduced $(v_g < v_r)$ as light spends more time in each microring before coupling to the next. In general, the group velocity can be further reduced by decreasing the coupling coefficient κ between the microrings. For the lossless CROW array of finite length, the power transmission reaches unity in the passband although there are also large ripples. These ripples are caused by reflections of the Bloch waves from the two terminated ends of the CROW array. For a finite CROW array, the ripples in the transmission response can be reduced by proper adjustment (or apodization) of the coupling coefficients. The analysis and design of CROW filters will be treated in Section 3.4.

Experimentally, CROW arrays of 100 micro-racetracks (Xia et al. 2007b) and 235 micro-racetracks (Cooper et al. 2010) have been demonstrated in the SOI material system for on-chip buffering applications. A major challenge in the realization of these long CROW arrays is that due to fabrication imperfections, the

---

[*] The transmission response of a finite CROW array is derived in Section 3.4.

microring resonators do not have identical resonant frequencies. This resonance mismatch causes high transmission loss and large ripples in the passband.

The analysis of 1D CROW arrays presented in this section can also be extended to obtain the dispersion relation of a 2D lattice of CMRs. In particular, it can be shown that the dispersion relation of an infinite 2D microring lattice can be expressed as a sum of two independent dispersion relations of 1D microring arrays in the two orthogonal directions (Chremmos and Uzunoglu 2008). Of more practical interest are the spectral responses of finite 2D microring lattices. In Section 3.7 we will develop methods for analyzing and designing microring filters of 2D coupling topologies.

3.2   Transfer Functions of Coupled Microring Optical Filters

Infinite periodic arrays of microring resonators are idealized structures since in practice the arrays must have finite lengths. However, truncation of an infinite microring array gives rise to surface states localized near the ends of the array, which manifest themselves as ripples in the transmission response. These ripples may also be understood as being caused by reflections due to impedance mismatch between the Bloch modes of the array and the excitation and transmission waves at the input and output ports. We can reduce or eliminate this impedance mismatch by apodizing the coupling coefficients in the array. The problem of determining a set of coupling coefficients that can generate a desired transmission response is called the microring filter synthesis problem.

In this chapter, we approach the analysis and design of microring optical filters from the point of view of classical filter theory. Specifically, the spectral response of a microring filter can be understood in terms of the poles and zeros of its transfer functions, and the design of the filter is accomplished by proper placement of its poles and zeros. Many well-known techniques for analog and digital filter design can be applied to the synthesis of microring filters. Thus, a familiarity with classical filter theory is essential for designing high-order microring filters. In this section, we will give a brief review of analog and digital filters and the general characteristics of their transfer functions. The remaining sections of the chapter will be devoted to the analysis and synthesis of important classes of microring filter architectures.

In general filters can be classified into two types: finite impulse response (FIR) and infinite impulse response (IIR) filters. As the names suggest, FIR filters are those whose impulse response function $h(t)$ lasts for a finite duration of time, whereas the impulse response of an IIR filter continues indefinitely. In integrated optics, FIR filters are commonly realized using directional couplers and Mach–Zehnder Interferometers (Madsen and Zhao 1999, Jinguji and Oguma 2000). IIR filters, on the other hand, are characterized by some feedback mechanisms and typically require a grating or a resonator. All microring filters are IIR filters.

The temporal response of an FIR or IIR filter may be continuous or discrete. A filter whose impulse response is a continuous function of time is called an analog filter. Its transfer function is given by the Laplace transform of the impulse response $h(t)$ and is expressed as a function of the complex frequency variable s. A filter whose response is a discrete function of time is called a digital filter. The impulse response of a digital filter is expressed as a time sequence $h(nT)$, where $n = 0, 1, 2, \ldots$, and T is a constant time interval or delay. The z-transform of the impulse response gives the transfer function of the digital filter, which is expressed as a function of the unit delay variable z.

In Chapter 2 we saw that the transfer function of a microring resonator obtained using the energy coupling formalism has the form of a transfer function of an analog filter in the s-domain. On the other hand, power coupling analysis of the device leads to a transfer function in the z-domain, where the unit delay variable z^{-1} represents the round-trip phase delay of the microring. In general, in the design of microring filters, it is more convenient to synthesize s-domain transfer functions using techniques based on the energy coupling formalism, whereas synthesis of z-domain transfer functions is more naturally performed in the framework of power coupling. Filter design methods based on energy coupling are also generally simpler than those based on power coupling. Another advantage of designing microring filters in the s-domain is that the transfer functions of a large class of well-known filters such as Butterworth, Chebyshev, elliptic, and linear phase filters are readily available. However, since the energy coupling formalism assumes weak coupling between the microrings, s-domain synthesis methods are strictly accurate only for narrowband filters whose passbands are much smaller than the FSR of the resonators. For broadband or strongly coupled microring filters, a more accurate design must be sought in the z-domain using the power coupling approach.

The transfer function describing the transmission response of an analog IIR filter can be expressed as a rational function in the s-domain,

$$H(s) = \frac{P(s)}{Q(s)} = \frac{K_0 \prod_{k=1}^{M} (s - z_k)}{\prod_{k=1}^{N} (s - p_k)}, \qquad (3.24)$$

where $P(s)$ and $Q(s)$ are polynomials of s, K_0 is a constant, and z_k and p_k are the zeros and poles, respectively, of the transfer function. The number of poles, N, gives the order of the filter. The number of zeros, M, is typically less than or at most equal to N. For passive filters, causality restricts the locations of the poles to the left half of the s-plane. Associated with the transfer function $H(s)$ is a complementary transfer function $F(s)$, which describes the reflection response of the filter. The reflection response $F(s)$ is related to the transmission response $H(s)$ via the Feldtkeller relation for a lossless network,

$$H(s)H^*(-s) + F(s)F^*(-s) = 1. \qquad (3.25)$$

Expressing $F(s)$ as a rational function of the form $F(s) = R(s)/Q(s)$, where $R(s)$ is a polynomial of at most degree N, we obtain from Equation 3.25

$$R(s)R^*(-s) = Q(s)Q^*(-s) - P(s)P^*(-s). \qquad (3.26)$$

By factoring the polynomial on the right-hand side of the above equation, we obtain two sets of roots, one lying in the left half of the s-plane and the other in the right half. Any combination of N roots taken from both sets can be used to form the polynomial $R(s)$, although it is common to choose the roots in the left half-plane. This choice of zeros yields what is known as the minimum phase reflection response.

In the inverse z-domain,[*] the transfer function of a digital IIR filter has the general form

---

[*] For convenience we will work in the inverse z-domain rather the z-domain in this book. Thus, for example, an Nth-degree polynomial $A(z^{-1})$ has the general form $A(z^{-1}) = a_0 + a_1 z^{-1} + \ldots + a_{N-1} z^{-(N-1)} + a_N z^{-N}$.

$$H(z^{-1}) = \frac{P(z^{-1})}{Q(z^{-1})} = \frac{K_0 \prod_{k=1}^{M} (z^{-1} - z_k)}{\prod_{k=1}^{N} (z^{-1} - p_k)}, \tag{3.27}$$

where $P(z^{-1})$ and $Q(z^{-1})$ are polynomials of z^{-1}, and z_k and p_k are the zeros and poles, respectively, in the inverse z-plane. For passive filters, the poles p_k are restricted to the region outside the unit circle in the inverse z-plane. The transmission response $H(z^{-1})$ and the complementary reflection response, $F(z^{-1}) = R(z^{-1})/Q(z^{-1})$, also satisfy the Feldtkeller relation

$$H(z^{-1})H^*(z) + F(z^{-1})F^*(z) = 1, \tag{3.28}$$

where $H^*(z)$ and $F^*(z)$ are the para-conjugates of the transfer functions. In terms of the polynomials P, Q, and R, the para-conjugate transfer functions can be expressed as $H^*(z) = \tilde{P}(z^{-1})/\tilde{Q}(z^{-1})$ and $F^*(z) = \tilde{R}(z^{-1})/\tilde{Q}(z^{-1})$, where $\tilde{P}(z^{-1})$, $\tilde{Q}(z^{-1})$, and $\tilde{R}(z^{-1})$ are the Hermitian conjugates[*] of the respective polynomials. Substituting these expressions into Equation 3.28, we obtain

$$R(z^{-1})\tilde{R}(z^{-1}) = Q(z^{-1})\tilde{Q}(z^{-1}) - P(z^{-1})\tilde{P}(z^{-1}). \tag{3.29}$$

The polynomial on the right-hand side of the above equation has two sets of roots, one lying inside the unit circle and the other lying outside. The polynomial $R(z^{-1})$ can be constructed from any combination of N of these roots. The choice of the roots lying outside the unit circle in the inverse z-plane gives the minimum phase filter design.

The design of a microring optical filter begins with the specification of the desired transfer functions H and F in either the s- or inverse z-domain.[†] The locations of the poles and zeros are chosen

---

[*] The Hermitian conjugate (or flip conjugate) of a polynomial $A(z^{-1})$ of degree N is defined as $\tilde{A}(z^{-1}) = z^{-N}A^*(z)$, where $A^*(z)$ is the para-conjugate (or para-Hermitian conjugate) of $A(z^{-1})$. The Hermitian conjugate can be obtained by reversing the coefficients of $A(z^{-1})$ and taking their complex conjugates. For example, if $A(z^{-1}) = a_0 + a_1 z^{-1} + \ldots + a_{N-1} z^{-(N-1)} + a_N z^{-N}$, then its Hermitian conjugate is $\tilde{A}(z^{-1}) = a_N^* + a_{N-1}^* z^{-1} + \ldots + a_1^* z^{-(N-1)} + a_0^* z^{-N}$.

[†] Typically one specifies the desired transmission characteristics of a filter, from which a suitable transmission transfer function $H(s)$ or $H(z^{-1})$ is constructed. Knowing $H(s)$ or $H(z^{-1})$, one can determine the complementary transfer function $F(s)$ or $F(z^{-1})$ for the reflection response using the Feldtkeller relation in Equation (3.26) or (3.29).

to achieve specific amplitude, phase, or group-delay characteristics of the filter. Given a set of prescribed spectral characteristics, the problem of determining a suitable transfer function is known as the filter approximation problem, for which a large body of literature exists in the field of analog and digital filter design (e.g., Lam 1979, Antoniou 1993, Ellis 1994). From the target transfer functions H and F, a suitable microring configuration and its coupling coefficients are then determined, which can reproduce the desired spectral response. This is called the filter synthesis problem. The remaining sections of this chapter are devoted to developing general techniques for analyzing and designing some of the most common microring filter architectures.

3.3   Cascaded All-Pass Microring Filters

Arrays of cascaded APMRs are the simplest microring architectures for realizing high-order all-pass optical filters. The device consists of N microring resonators side coupled to the same bus waveguide as shown in Figure 3.8. A light signal s_i applied to the input port couples sequentially to all the microrings. The transmitted signal s_t is the result of the interference between the signals from all the microrings with the input signal. If there is no intrinsic loss in the resonators, all the input power will appear at the output port. Signals at all frequencies will thus pass through the structure unattenuated but can acquire a complex phase response. Cascaded all-pass microring arrays have important applications as phase filters, dispersion compensators, and optical delay lines.

To determine the general transfer function of a cascaded all-pass microring filter, we consider the array of N APMRs shown in Figure 3.8. Each microring k has radius R_k and is coupled to the common bus waveguide via field coupling coefficient κ_k (transmission coefficient τ_k). For simplicity we assume that all the microrings have the same round-trip amplitude attenuation a_{rt}. We denote the round-trip phase of microring k as $\phi_k = \phi_0 + \Delta\phi_k$, where ϕ_0 is a reference round-trip phase (at a center wavelength λ_0) and $\Delta\phi_k$ is the phase

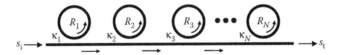

Figure 3.8   Schematic of an array of N cascaded APMRs.

detune of microring k with respect to ϕ_0. The transfer function of the array is simply the product of the transfer functions of individual APMRs. Using the transfer function of an APMR in Equation 2.44, we can write the transfer function of the APMR array as

$$H(z^{-1}) = \frac{s_t}{s_i} = \prod_{k=1}^{N} \frac{\tau_k - a_{rt}\exp(-j\Delta\phi_k)z^{-1}}{1 - \tau_k a_{rt}\exp(-j\Delta\phi_k)z^{-1}}, \qquad (3.30)$$

where $z^{-1} = e^{-j\phi_0}$. Equation 3.30 has the form of the transfer function of an Nth-order all-pass filter with poles $p_k = (1/\tau_k a_{rt})\exp(j\Delta\phi_k)$ and zeros $z_k = (\tau_k/a_{rt})\exp(j\Delta\phi_k)$. Each microring in the array is responsible for generating a pole and a zero whose locations are determined by the transmission coefficient τ_k and the microring phase detune $\Delta\phi_k$. In the absence of loss, the poles and zeros are related simply by $z_k = 1/p_k^*$.

The phase response ψ of the filter is the sum of the phases of the N APMRs. Using Equation 2.59 for the phase response ψ_k of all-pass microring k, we get

$$\psi = \sum_{k=1}^{N} \psi_k = \sum_{k=1}^{N} \tan^{-1}\left(\frac{a_{rt}\sin\phi_k}{\tau_k - a_{rt}\cos\phi_k}\right) - \sum_{k=1}^{N} \tan^{-1}\left(\frac{\tau_k a_{rt}\sin\phi_k}{1 - \tau_k a_{rt}\cos\phi_k}\right).$$

$$(3.31)$$

Similarly, the group-delay response of the array can be obtained by summing up the group delays of the individual APMRs. Using Equation 2.62 for the group delay of an APMR, we obtain the group-delay response of the array as follows:

$$\tau_g = \sum_{k=1}^{N} \frac{S_{p,k}T_{rt,k}}{1 + F_k\sin^2(\phi_k/2)} - \sum_{k=1}^{N} \frac{S_{z,k}T_{rt,k}}{1 + G_k\sin^2(\phi_k/2)}. \qquad (3.32)$$

In the above equation, $T_{rt,k}$ is the round-trip delay time of microring k, and $S_{p,k}$ and $S_{z,k}$ are the peak group delays due to the poles and zeros, respectively,

$$S_{p,k} = \frac{1}{2}\left(\frac{1 + \tau_k a_{rt}}{1 - \tau_k a_{rt}}\right), \qquad (3.33)$$

$$S_{z,k} = \frac{1}{2}\left(\frac{\tau_k + a_{rt}}{\tau_k - a_{rt}}\right), \qquad (3.34)$$

and F_k and G_k are, respectively, given by

$$F_k = 4\tau_k a_{\mathrm{rt}} / (1 - \tau_k a_{\mathrm{rt}})^2 , \tag{3.35}$$

$$G_k = 4\tau_k a_{\mathrm{rt}} / (\tau_k - a_{\mathrm{rt}})^2 . \tag{3.36}$$

If all N microrings are identical with identical coupling coefficients, the phase and group-delay responses of the array are simply N times those of a single APMR. Such an array can be used as an optical delay element whose group delay is enhanced by N times over that of a single APMR.

The cascaded APMR array can be used to synthesize a given all-pass transfer function of order N by selecting the coupling coefficients and phase shifts of the microrings to realize the required poles and zeros. As an example, we consider the design of an all-pass Bessel filter which has a maximally flat group delay response. In the inverse z-domain, the transfer function of an all-pass Bessel filter of order N is given by (Prabhu and Van 2008)

$$H_{\mathrm{ap}}(z^{-1}) = \frac{\prod_{k=1}^{N} \left(1 - z^{-1} e^{-r_k T_{\mathrm{rt}}}\right)}{\prod_{k=1}^{N} \left(z^{-1} - e^{-r_k T_{\mathrm{rt}}}\right)}, \tag{3.37}$$

where r_k are the roots of the Bessel polynomial of degree N. The poles and zeros of the filter are located at $p_k = \exp(-r_k T_{\mathrm{rt}})$ and $z_k = \exp(r_k T_{\mathrm{rt}})$, respectively. Assuming that the microrings have very low loss ($a_{\mathrm{rt}} \approx 1$), the transmission coefficient and phase detune of microring k can be determined from $\tau_k = 1/|p_k|$ and $\Delta\phi_k = \angle p_k$, respectively. For a fifth-order Bessel filter with a 50 GHz bandwidth, the poles of the filter are located at $p_k = \{1.1045 \pm j0.1995, 1.1780 \pm j0.1036, 1.2006\}$. Assuming that the microrings have an FSR of 1 THz, we compute the transmission coefficients and phase detunes to be $\tau_k = \{0.891, 0.891, 0.846, 0.846, 0.833\}$ and $\Delta\phi_k = \{0.179, -0.179, 0.088, -0.088, 0\}$. Figure 3.9 shows the power transmission and group-delay responses of the filter in the presence of 1% power loss in the resonators. We see that the group-delay response has a constant value of 40 ps over the 50 GHz filter bandwidth. The transmission response is also flat over this bandwidth, with an insertion loss of -1.75 dB due to loss in the microrings.

Another important application of cascaded APMR arrays is the realization of dispersion compensators (Madsen and Lenz 1998).

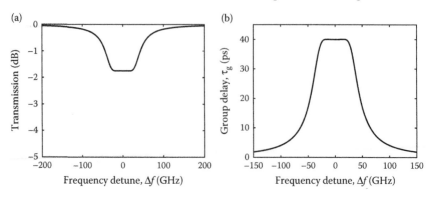

Figure 3.9 Spectral responses of (a) power transmission and (b) group delay of a fifth-order all-pass microring Bessel filter with maximally flat group delay.

These devices are typically designed to provide a constant chromatic dispersion D over a bandwidth B. To achieve a group-delay response with linear slope D, we can use a series of N all-pass microrings with resonances distributed over the bandwidth B. By designing the microrings to have increasing or decreasing peak group delays, positive or negative linear dispersion slopes can be obtained. Figure 3.10 shows an example of a dispersion compensator consisting of five cascaded APMRs. The phase detunes of the microrings are fixed at the values $\Delta\phi_k = \{-0.45\pi, -0.2\pi, 0, 0.15\pi, 0.25\pi\}$ and the transmission coefficients are set to be $\tau_k = \{0.333, 0.429, 0.500, 0.556, 0.600\}$, which are

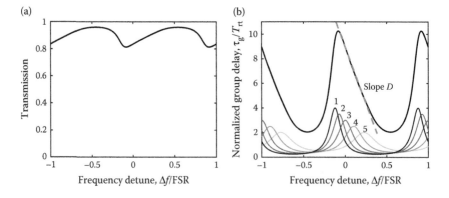

Figure 3.10 Spectral responses of (a) power transmission and (b) normalized group delay (τ_g/T_{rt}) of a dispersion compensator consisting of five cascaded APMRs. The curves labeled from 1 to 5 are the group delay responses of individual microring resonators.

numerically optimized to give a linear dispersion slope in the group delay. The microring resonators are assumed to have a round-trip loss of 2%, which causes a small ripple of about 0.7 dB in the transmission response of the array (Figure 3.10a). We also see in Figure 3.10b that the group-delay response exhibits a linear dispersion slope of $D = 18T_{rt}/\Delta\lambda_{FSR}$ over a bandwidth close to half a microring FSR. For example, if the microrings are designed to have an FSR of 25 GHz ($\Delta\lambda_{FSR} = 0.2$ nm and round-trip time $T_{rt} = 40$ ps), the chromatic dispersion D of the array would be 3.6 ns/nm. Larger values of the dispersion slope can be obtained by using more microring resonators. The slope D can also be made negative by simply reversing the sign of each phase detune value Df_k. A tunable dispersion compensator based on a similar cascaded APMR array has been experimentally demonstrated using Ge-doped silica microrings in Madsen et al. (1999).

3.4   Serially Coupled Microring Filters

Serially coupled microring filters (or CROW filters) are the most common type of high-order microring filters since these devices are relatively simple to design and fabricate. Microring filters of various orders have been demonstrated in a variety of material systems including GaAs/AlGaAs (Hryniewicz et al. 2000), SiON (Little et al. 2004), SiN (Barwicz et al. 2004), and SOI (Xia et al. 2007a). In this section, we will first develop the theory and techniques for analyzing and designing serially coupled microring filters using the energy coupling formalism. This will be followed by a formulation of the device transfer functions in terms of power coupling, which is more accurate for broadband or strongly coupled microring devices. A method for synthesizing serially coupled microring filters using the power coupling formalism will also be developed.

3.4.1   Energy coupling analysis of serially coupled microring filters

We consider an array of N serially coupled microring resonators as shown in Figure 3.11. Each microring i has resonant frequency ω_i, which may be slightly detuned from a center frequency ω_0 (typically taken to be the center frequency of the filter passband). If the microrings are of approximately the same size, we may assume that they have the same intrinsic loss. We denote the wave amplitude in microring i as a_i, which is normalized with respect to energy so that $|a_i|^2$ gives the energy stored in the microring. Coupling between

Input

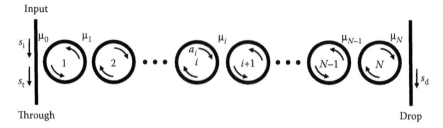

Through   Drop

Figure 3.11   Schematic of a serially coupled microring filter consisting of *N* microring resonators.

two adjacent microrings i and $i+1$ is denoted by the ring-to-ring energy coupling coefficient μ_i ($1 \leq i \leq N{-}1$). In addition, microrings 1 and N are also coupled to an input waveguide and an output wave-guide via ring-to-bus coupling coefficients μ_0 and μ_N, respectively. An input signal s_i carrying power $|s_i|^2$ is applied to the input port which supplies energy to microring 1 at a rate of $\mu_0^2 |s_i|^2$. At the same time, energies in microrings 1 and N are coupled out to the input and output bus waveguides, respectively, and transmitted to the through port and drop port as power-normalized waves s_t and s_d.

In each microring resonator the rate of change of energy is due to a combination of intrinsic loss and coupling to neighbor microrings, as well as coupling to the bus waveguides in the case of microrings 1 and N. We can thus write the coupled mode equations describing energy transfer between the microrings as (Haus 1984, Little et al. 1997)

$$\frac{da_1}{dt} = (j\omega_1 - \gamma_i - \gamma_0)a_1 - j\mu_1 a_2 - j\mu_0 s_i,$$

$$\frac{da_2}{dt} = (j\omega_2 - \gamma_i)a_2 - j\mu_1 a_1 - j\mu_2 a_3,$$

$$\cdots \tag{3.38}$$

$$\frac{da_{N-1}}{dt} = (j\omega_{N-1} - \gamma_i)a_{N-1} - j\mu_{N-2}a_{N-2} - j\mu_{N-1}a_N,$$

$$\frac{da_N}{dt} = (j\omega_N - \gamma_i - \gamma_N)a - j\mu_{N-1}a_{N-1}.$$

In the above equations, γ_i is the amplitude decay rate due to intrinsic loss, which is related to the propagation loss α in the microring waveguide by $\gamma_i = \alpha v_g / 2$, where v_g is the group velocity. The decay rates γ_0 and γ_N are due to coupling from microrings 1 and N

to the input and output bus waveguides, respectively, and can be computed from the ring-to-bus energy coupling coefficients using Equation 2.73 as $\gamma_0 = \mu_0^2/2$ and $\gamma_N = \mu_N^2/2$.

In microring filter design, it is necessary to relate the energy coupling coefficients to the field coupling coefficients, which are then used to determine the coupling gaps between the microrings or between the microrings and the bus waveguides. The ring-to-bus energy coupling coefficients can be computed from the field coupling coefficients κ_0 and κ_N using Equation 2.76,

$$\mu_i = \kappa_i\sqrt{\frac{v_g}{2\pi R_i}} = \frac{\kappa_i}{\sqrt{T_{rt,i}}}, \quad i = 0, N, \tag{3.39}$$

where R_i is the radius of microring i and $T_{rt,i}$ is its round-trip time. To relate the ring-to-ring energy coupling coefficients μ_i $(1 \leq i \leq N-1)$ to the corresponding field coupling coefficients κ_i, we observe that the rate of energy coupling from ring $i+1$ to ring i is given by

$$\mu_i^2 |a_{i+1}(t)|^2 = \mu_i^2 \left(\frac{2\pi R_{i+1}}{v_g}\right) |A_{i+1}(t)|^2 \equiv g_i^2 |A_{i+1}(t)|^2, \tag{3.40}$$

where $|A_{i+1}(t)|^2$ is the power circulating in microring $i+1$, and $g_i^2 |A_{i+1}|^2$ is the rate at which the power-normalized wave A_{i+1} supplies energy to microring i. We can regard the wave A_{i+1} as playing the same role as the power-normalized wave s_i, which supplies energy to microring 1 at a rate equal to $\mu_0^2 |s_i|^2$. Using Equation 3.39, we can compute g_i from the field coupling coefficient κ_i as $g_i = \kappa_i\sqrt{v_g/2\pi R_i}$. We thus have

$$g_i^2 = \mu_i^2(2\pi R_{i+1}/v_g) = \kappa_i^2(v_g/2\pi R_i),$$

which yields

$$\mu_i = \kappa_i \frac{v_g}{2\pi\sqrt{R_i R_{i+1}}} = \frac{\kappa_i}{\sqrt{T_{rt,i} T_{rt,i+1}}}, \quad 1 \leq i \leq N-1. \tag{3.41}$$

If the microrings are identical with the same round-trip time, the above equation simplifies to $\mu_i = \kappa_i/T_{rt}$.

To obtain the transfer functions of the serially coupled microring filter, we solve the coupled mode equations in Equation 3.38 for the case of a harmonic input wave excitation of the form $s_i \sim e^{j\omega t}$. In this case the solutions for the wave amplitudes a_i in the microrings will also vary as $a_i \sim e^{j\omega t}$. Applying these solutions to Equation 3.38, we get

$$[j(\omega - \omega_1) + \gamma_i + \gamma_0]a_1 + j\mu_1 a_2 = -j\mu_0 s_i,$$
$$j\mu_1 a_1 + [j(\omega - \omega_2) + \gamma_i]a_2 + j\mu_2 a_3 = 0,$$

$$\cdots \tag{3.42}$$

$$j\mu_{N-2}a_{N-2} + [j(\omega - \omega_{N-1}) + \gamma_i]a_{N-1} + j\mu_{N-1}a_N = 0,$$
$$j\mu_{N-1}a_{N-1} + [j(\omega - \omega_N) + \gamma_i + \gamma_N]a_N = 0.$$

It is convenient to write the terms $\omega - \omega_i$ in the above equations as

$$\omega - \omega_i = (\omega - \omega_0) - (\omega_i - \omega_0) = \Delta\omega - \delta\omega_i, \tag{3.43}$$

where $\Delta\omega$ is the frequency detune from the center frequency ω_0 and $\delta\omega_i$ is the deviation of the resonance of microring i from ω_0. Defining the complex frequency $s = j\Delta\omega + \gamma_i$, we can express Equation 3.42 in the matrix form

$$\mathbf{K}a = s, \tag{3.44}$$

where $a = [a_1, a_2, \ldots a_{N-1}, a_N]^T$ is the wave amplitude array, $s = [-j\mu_0 s_i, 0, \ldots 0, 0]^T$ is the input excitation vector, and $\mathbf{K}$ is a symmetric tridiagonal matrix,

$$\mathbf{K} = \begin{bmatrix} s - j\delta\omega_1 + \gamma_0 & j\mu_1 & & & & \\ j\mu_1 & s - j\delta\omega_2 & j\mu_2 & & & \\ & \cdot & \cdot & \cdot & & \\ & & j\mu_{N-2} & s - j\delta\omega_{N-1} & j\mu_{N-1} & \\ & & & j\mu_{N-1} & s - j\delta\omega_N + \gamma_N \end{bmatrix} \tag{3.45}$$

For a given input wave amplitude s_i, the solution of the matrix Equation 3.44 gives the wave amplitudes a_i in the microrings.

At the drop port of the microring filter, the transmitted signal is related to the wave amplitude in microring N by $s_d = -j\mu_N a_N$. We can thus write the drop port transfer function of the filter as

$$H_d = \frac{s_d}{s_i} = \frac{-j\mu_N a_N}{s_i}. \tag{3.46}$$

From Equation 3.44, we obtain the solution for a_N as $a_N = \det(\mathbf{A})/\det(\mathbf{K})$, where $\mathbf{A}$ is the matrix obtained by replacing the last column of $\mathbf{K}$ with s. The determinant of $\mathbf{A}$ can be explicitly computed as follows:

$$\det(\mathbf{A}) = -j\mu_0 s_i \prod_{k=1}^{N-1}(-j\mu_k) = (-j)^N(\mu_0\mu_1\mu_2 \cdots \mu_{N-1})s_i. \qquad (3.47)$$

For the tridiagonal matrix $\mathbf{K}$, its determinant is given by $\det(\mathbf{K}) = C_N$, where C_N is the Nth-continuant obtained from the recursive formula*

$$_*C_k = K_{N-K+1,N-K+1}C_{k-1} + \mu_{N-K+1}^2 C_{k-2}, \quad (k \geq 2) \qquad (3.48)$$

with $C_0 = 1$ and $C_1 = K_{N,N}$. The drop port transfer function of the microring filter can thus be expressed as

$$H_d(s) = \frac{(-j)^{N+1}(\mu_0\mu_1\mu_2 \cdots \mu_N)}{C_N(s)}, \qquad (3.49)$$

where the continuant $C_N(s)$ has the form of a polynomial of degree N.

To determine the transfer function at the through port of the filter, we observe that the through port signal is given by $s_t = s_i - j\mu_0 a_1$. We can thus write the through port transfer function as

$$H_t = \frac{s_t}{s_i} = 1 - \frac{j\mu_0 a_1}{s_i}. \qquad (3.50)$$

The wave amplitude in microring 1 can be obtained from $a_1 = \det(\mathbf{B})/\det(\mathbf{K})$, where $\mathbf{B}$ is the matrix obtained by replacing the first column of $\mathbf{K}$ with s. The solution can be expressed in terms of the continuants as

$$a_1 = -j\mu_0 s_i \frac{C_{N-1}(s)}{C_N(s)}, \qquad (3.51)$$

which may also be written in the form of a continued fraction,

$$a_1 = \cfrac{-j\mu_0 s_i}{s - j\delta\omega_1 + \gamma_0 + \cfrac{\mu_1^2}{s - j\delta\omega_2 + \cfrac{\mu_2^2}{s - j\delta\omega_3 + \ldots \cfrac{\mu_{N-1}^2}{s - j\delta\omega_N + \gamma_N}}}}. \qquad (3.52)$$

---

* Equation 3.48 expresses the continuant for the matrix $\mathbf{K}$ whose columns and rows have been flipped left-to-right and up-to-down, respectively. This allows the same recursive formula to be used to calculate the determinant of the matrix $\mathbf{B}$ in Equation 3.51.

The above solution for a_1 can also be obtained through a backward substitution procedure of the system $\mathbf{K}a = s$. Using the result in Equation 3.52, we obtain the transfer function at the through port as

$$H_t(s) = 1 - \cfrac{\mu_0^2}{s - j\delta\omega_1 + \gamma_0 + \cfrac{\mu_1^2}{s - j\delta\omega_2 + \cfrac{\mu_2^2}{s - j\delta\omega_3 + \ldots \cfrac{\mu_{N-1}^2}{s - j\delta\omega_N + \gamma_N}}}}.$$

(3.53)

Equations 3.49 and 3.53 give the closed-form expressions for the transfer functions at the drop port and through port of an Nth-order serially coupled microring filter. We note that the drop port transfer function has N poles but no finite zeros. Thus, the serial microring coupling configuration can only be used to realize all-poles transfer functions, such as those of Butterworth (or maximally flat) filters, Chebyshev filters, and Bessel filters.

If the output bus waveguide is removed from the microring array in Figure 3.11, the structure becomes an Nth-order all-pass microring filter. The transfer function $H_t = s_t/s_i$ of the device is given by Equation 3.53 with the output bus-to-ring coupling coefficient μ_N set to 0 (so that $\gamma_N = 0$). In the all-pass configuration, the serially coupled microring filter is equivalent to the cascaded all-pass microring array in Section 3.3 and can be used to realize all-pass transfer functions of any order N.

3.4.2 Energy coupling synthesis of serially coupled microring filters

In this section we develop a simple method for synthesizing serially coupled microring filters to achieve a target spectral response. The method exploits the fact that the through port transfer function $H_t(s)$ of the microring filter can be expressed as a continued fraction (Prabhu et al. 2008a). Suppose that the target filter has an Nth-order through port (or reflection) transfer function given by

$$H_t(s) = \frac{R(s)}{Q(s)} = \frac{s^N + r_{N-1}s^{N-1} + \cdots + r_1 s + r_0}{s^N + q_{N-1}s^{N-1} + \cdots + q_1 s + q_0}.$$

(3.54)

We form the expression

$$1 - H_t(s) = \frac{Q(s) - R(s)}{Q(s)} = (q_{N-1} - r_{N-1}) \frac{M_{N-1}(s)}{Q(s)}, \qquad (3.55)$$

where $M_{N-1}(s)$ is a polynomial of degree $N-1$ with the leading coefficient equal to 1. Assuming that the microrings are synchronously tuned $(\delta\omega_i = 0)$ and lossless, we rearrange Equation 3.53 to read

$$1 - H_t(s) = \cfrac{\mu_0^2}{(s + \gamma_0) + \cfrac{\mu_1^2}{s + \cfrac{\mu_2^2}{s + \cdots \cfrac{\mu_{N-1}^2}{s + \gamma_N}}}}, \qquad (3.56)$$

where $s = j\Delta\omega$. Comparing Equations 3.55 and 3.56 and noting that the leading coefficients of both $M_{N-1}(s)$ and $Q(s)$ are 1, we obtain $\mu_0^2 = q_{N-1} - r_{N-1}$. By expressing $M_{N-1}(s)/Q(s)$ in the form of a continued fraction, we can determine the remaining coupling coefficients μ_k. Specifically, carrying out the division

$$\frac{Q(s)}{M_{N-1}(s)} = (s + \gamma_0) + \frac{b \cdot M_{N-2}(s)}{M_{N-1}(s)} \qquad (3.57)$$

where $\gamma_0 = \mu_0^2 / 2$ and $M_{N-2}(s)$ are polynomial of degree $N-1$, we obtain $b = \mu_1^2$. Next, by dividing $M_{N-1}(s)$ by $M_{N-2}(s)$, we can get μ_2. This process is repeated until all the coupling coefficients are determined. This method can also be used to synthesize all-pass microring filters consisting of N serially coupled microring resonators.

Two common types of filters that can be realized with the serial coupling configuration are the Butterworth and Chebyshev filters. For these filters, closed-form design formulas for computing the energy coupling coefficients can be obtained. Butterworth filters (also called maximally flat filters) of order N have the property that the first $(2N-1)$ derivatives of the frequency response $|H(j\omega)|^2$ are zero at the center frequency. Chebyshev filters provide steeper skirt roll-off than Butterworth filters but have equi-ripples in the passband. The ripples are characterized by a small positive parameter ε such that $|H(j\omega)|^2$ oscillates between 1 and $1/(1 + \varepsilon^2)$ within the passband. The transfer function of an Nth-order Butterworth or Chebyshev filter can be expressed as

$$H(s) = \frac{1}{K_0} \prod_{k=1}^{N} \frac{1}{s - p_k},$$

(3.58)

where $s = 2j(\omega - \omega_0)/\Delta\omega_B$ and $\Delta\omega_B$ is the filter bandwidth. For Butterworth filters, the constant K_0 is equal to 1 and $\Delta\omega_B$ is the 3 dB bandwidth. For Chebyshev filters, $K_0 = 2^{N-1}\varepsilon$ and $\Delta\omega_B$ is the ripple bandwidth. The poles p_k of the filter are determined by

$$p_k = -a\sin\theta_k + jb\cos\theta_k,$$
$$\theta_k = \frac{(2k-1)\pi}{2N}.$$

(3.59)

For Butterworth filters, $a = 1$ and $b = 1$, whereas for Chebyshev filters they are given by

$$a = \sinh\left[\frac{1}{N}\sinh^{-1}\left(\frac{1}{\varepsilon}\right)\right],$$
$$b = \cosh\left[\frac{1}{N}\sinh^{-1}\left(\frac{1}{\varepsilon}\right)\right].$$

(3.60)

Butterworth and Chebyshev filters of order N can be realized using N serially coupled microring resonators. The energy coupling coefficients can be computed directly from the formulas (Van 2006)

$$\mu_0^2 = \mu_N^2 = \frac{c_0\Delta\omega_B/2}{\sin(\pi/2N)},$$
$$\mu_k^2 = \frac{(c_k\Delta\omega_B/4)^2}{\sin\left[(2k-1)\pi/2N\right]\sin\left[(2k+1/2N)\pi\right]}, \quad 1 \le k \le N-1$$

(3.61)

In the above expressions, $c_k = 1$ for Butterworth filters and

$$c_k^2 = \sin^2\left(\frac{k\pi}{N}\right) + \sinh^2\left[\frac{1}{2N}\ln\left(\frac{\sqrt{1+\varepsilon^2}+1}{\sqrt{1+\varepsilon^2}-1}\right)\right], \quad 0 \le k \le N$$

(3.62)

for Chebyshev filters with ripple parameter ε.

As an example, we consider the design of a fourth-order Butterworth filter and a Chebyshev filter of the same order with a

0.5 dB ripple. The filter bandwidth is specified to be $\Delta f_B = 50$ GHz. Using Equation 3.61, we compute the energy coupling coefficients to get $\mu_0 = \mu_4 = 20.26$ GHz$^{1/2}$, $\{\mu_1, \mu_2, \mu_3\} = \{132.09, 85.01, 132.09\}$ GHz for the Butterworth filter, and $\mu_0 = \mu_4 = 13.71$ GHz$^{1/2}$, $\{\mu_1, \mu_2, \mu_3\} = \{111.30, 93.51, 111.30\}$ GHz for the Chebyshev filter. Using microrings with FSR = 1 THz, we calculate the corresponding field coupling coefficients using Equations 3.39 and 3.41 to be $\{\kappa_0, \kappa_1, \kappa_2, \kappa_3, \kappa_4\} = \{0.641, 0.132, 0.085, 0.132, 0.641\}$ for the Butterworth filter and $\{\kappa_0, \kappa_1, \kappa_2, \kappa_3, \kappa_4\} = \{0.434, 0.111, 0.094, 0.111, 0.434\}$ for the Chebyshev filter. In Figure 3.12 we plot the spectral responses at the drop port and through port of both filters. The Chebyshev filter is seen to have a sharper roll-off than the Butterworth filter at the expense of a small ripple in the passband. In general, the roll-off can be made steeper at the expense of increased in-band ripples.

Due to their flat-top passbands and relatively steep roll-offs, high-order Butterworth microring filters are especially of interest for applications as add-drop filters in WDM communication systems. Filters of various orders have been fabricated in a variety of material systems. One of the earliest experimental results that demonstrated the viability of these devices is shown in Figure 3.13. The plot shows the measured and theoretical spectral responses of Butterworth microring filters of increasing orders from 1 to 6 fabricated in the Hydex material system (Van et al. 2004). The microrings had a radius of 42 μm and the filters were designed to have a 3 dB bandwidth of around 100 GHz. Fairly good agreement between the measured and designed responses can be seen. It is also evident

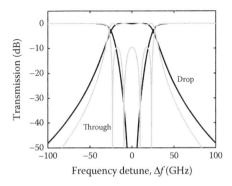

Figure 3.12 Spectral responses of fourth-order serially coupled microring filters: Butterworth filters (black lines) and Chebyshev filters (gray lines).

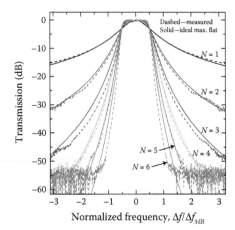

Figure 3.13 Measured and theoretical spectral responses of Butterworth microring filters of orders 1–6 fabricated in the Hydex material. (From Van, V., et al., 2004. Micro-ring resonator filters. In *The 17th Annual Meeting of the IEEE Lasers and Electro-Optics Society*, Vol. 2, pp. 571–572.)

that the filter passband becomes flatter and the roll-off becomes steeper as the filter order is increased.

3.4.3 Power coupling analysis of serially coupled microring filters

The energy coupling methods developed in Sections 3.4.1 and 3.4.2 are strictly valid only for narrowband microring filters, that is, filters whose 3 dB bandwidths are much smaller than the FSRs of the microrings. For broadband or, equivalently, strongly coupled microring filters, a more accurate analysis or design must be performed using the power coupling formalism. In this approach, the microring device is treated as a digital IIR filter, whose transfer functions are specified in terms of the delay variable z^{-1}.

In this section, we apply the transfer matrix method developed in Section 3.1.3 for infinite CROW arrays to obtain the z-domain transfer functions of a serially coupled microring filter of order N (Orta et al. 1995, Poon et al. 2003). For simplicity, we assume that the microrings are identical with the same radius R, propagation constant β, and loss α in the microring waveguides. Coupling between adjacent microrings i and $i+1$ is denoted by the field coupling coefficient κ_i $(1 \leq i \leq N-1)$. Microrings 1 and N are also coupled to the input and output bus waveguides via field coupling coefficients κ_0 and κ_N, respectively.

In each microring i, we label the fields a_i, b_i, c_i, d_i in the direction of wave propagation, as shown in Figure 3.14. The four fields are related by a transfer matrix $\mathbf{P}$ defined as

$$\begin{bmatrix} c_i \\ d_i \end{bmatrix} = \begin{bmatrix} 0 & a_{rt}^{1/2}e^{-j\phi_{rt}/2} \\ a_{rt}^{-1/2}e^{j\phi_{rt}/2} & 0 \end{bmatrix} \begin{bmatrix} a_i \\ b_i \end{bmatrix} = z^{1/2}\begin{bmatrix} 0 & z^{-1} \\ 1 & 0 \end{bmatrix}\begin{bmatrix} a_i \\ b_i \end{bmatrix} \equiv z^{1/2}\mathbf{P}\begin{bmatrix} a_i \\ b_i \end{bmatrix},$$

(3.63)

where $z^{-1} = a_{rt}e^{-j\phi_{rt}}$, $\phi_{rt} = 2\pi\beta R$ is the round-trip phase and $a_{rt} = e^{-\pi\alpha R}$ is the round-trip attenuation factor in the microrings. The coupling junction between microrings i and $i{+}1$ is described by the transfer matrix $\mathbf{K}_i$ given in Equation 3.21,

$$\begin{bmatrix} a_{i+1} \\ b_{i+1} \end{bmatrix} = \frac{1}{j\kappa_i}\begin{bmatrix} \tau_i & -1 \\ 1 & -\tau_i \end{bmatrix}\begin{bmatrix} c_i \\ d_i \end{bmatrix} \equiv \mathbf{K}_i\begin{bmatrix} c_i \\ d_i \end{bmatrix},$$

(3.64)

where $\tau_i = \sqrt{1-\kappa_i^2}$. Combining Equations 3.63 and 3.64, we obtain the transfer matrix for each unit element i in the microring array as follows:

$$\begin{bmatrix} a_{i+1} \\ b_{i+1} \end{bmatrix} = z^{1/2}\mathbf{K}_i\mathbf{P}\begin{bmatrix} a_i \\ b_i \end{bmatrix}.$$

(3.65)

The transfer matrix $\mathbf{T}$ of the array of N microrings is then

$$\begin{bmatrix} s_a \\ s_d \end{bmatrix} = z^{N/2}(\mathbf{K}_N\mathbf{P})(\mathbf{K}_{N-1}\mathbf{P})\cdots(\mathbf{K}_1\mathbf{P})\mathbf{K}_0\begin{bmatrix} s_i \\ s_t \end{bmatrix} \equiv z^{N/2}\mathbf{T}\begin{bmatrix} s_i \\ s_t \end{bmatrix}.$$

(3.66)

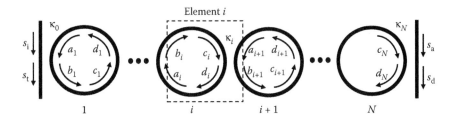

Figure 3.14 Schematic of N serially coupled microrings for transfer matrix analysis.

The transfer functions at the drop port and through port of the microring filter are obtained by setting $s_a = 0$ in Equation 3.66 and solving for s_t/s_i and s_d/s_i. The results are

$$H_t(z) = \frac{s_t}{s_i} = -\frac{T_{11}}{T_{12}}, \tag{3.67}$$

$$H_d(z) = \frac{s_d}{s_i} = z^{N/2}\left(T_{21} - \frac{T_{11}T_{22}}{T_{12}}\right) = -z^{N/2}\frac{\det(\mathbf{T})}{T_{12}}. \tag{3.68}$$

The determinant of $\mathbf{T}$ is just the product of the determinants of the matrices $\mathbf{K}_i$ and $\mathbf{P}$ in Equations 3.63 and 3.64. Since $\det(\mathbf{P}) = -z^{-1}$ and $\det(\mathbf{K}_i) = -1$, we obtain $\det(\mathbf{T}) = -z^{-N}$. With this result, we can write the transfer function at the drop port as

$$H_d(z) = \frac{z^{-N/2}}{T_{12}}. \tag{3.69}$$

If we define the transfer matrix $\mathbf{T}_k$ of the first k microrings as

$$\mathbf{T}_k = (\mathbf{K}_k\mathbf{P})(\mathbf{K}_{k-1}\mathbf{P})\cdots(\mathbf{K}_1\mathbf{P})\mathbf{K}_0, \tag{3.70}$$

then we have the recursive formula, $\mathbf{T}_k = (\mathbf{K}_k\mathbf{P})\mathbf{T}_{k-1}$. Starting from $\mathbf{T}_0 = \mathbf{K}_0$ and applying the recursive formula, we find that the transfer matrix $\mathbf{T} = \mathbf{T}_N$ of N serially coupled microrings has the general form

$$\mathbf{T}_N = \frac{1}{(j\kappa_0)(j\kappa_1)\cdots(j\kappa_N)}\begin{bmatrix} R_N(z^{-1}) & \sigma_N Q_N(z^{-1}) \\ \tilde{Q}_N(z^{-1}) & \sigma_N\tilde{R}_N(z^{-1}) \end{bmatrix}, \tag{3.71}$$

where R_N and Q_N are polynomials of degree N, $\tilde{R}_N$, and $\tilde{Q}_N$ are their Hermitian conjugates and $\sigma_N = +/-1$ if N is odd/even. The polynomial Q_N has the leading coefficient equal to 1 (coefficient of the z^{-N} term). The polynomials R_N and Q_N obey the recursive relations

$$R_k(z^{-1}) = \tau_k z^{-1}\tilde{Q}_{k-1}(z^{-1}) - R_{k-1}(z^{-1}), \tag{3.72}$$

$$Q_k(z^{-1}) = Q_{k-1}(z^{-1}) - \tau_k z^{-1}\tilde{R}_{k-1}(z^{-1}), \tag{3.73}$$

for $1 \leq k \leq N$ with $R_0 = \tau_0$ and $Q_0 = 1$. Using Equations 3.67 and 3.69, we can write the transfer functions of a serially coupled microring filter of order N as

$$H_d(z^{-1}) = \sigma_N z^{-N/2} \frac{(j\kappa_0)(j\kappa_1)\cdots(j\kappa_N)}{Q_N(z^{-1})}, \qquad (3.74)$$

$$H_t(z^{-1}) = -\frac{R_N(z^{-1})}{Q_N(z^{-1})}. \qquad (3.75)$$

The above equations show that the drop port and through port transfer functions of the filter have N poles, which are given by the roots of the polynomial Q_N. The drop port has no transmission zeros (other than the trivial zeros at $z^{-1} = 0$), while the through port has N zeros given by the roots of R_N.

3.4.4 Power coupling synthesis of serially coupled microring filters

Given the transfer functions of an all-poles filter in the inverse z-domain, it is possible to determine the coupling coefficients of a serially coupled microring filter that will reproduce the desired spectral responses. The filter synthesis method we develop here is similar to the technique first introduced in Orta et al. (1995) and is based on the transfer matrix method and order reduction technique. In the method, we start from the Nth-order transmission and reflection transfer functions of the target filter and extract the coupling coefficients of the unit elements (microrings) in the array one by one starting with the last coefficient (κ_N). Each time a coupling coefficient is extracted, the order of the transfer functions is reduced by one (i.e., the array is reduced by one unit element), and the procedure is repeated until the first coefficient (κ_0) is determined.

Suppose we know the polynomials R_k and Q_k of the transfer matrix $\mathbf{T}_k$ of the first k microrings. From the recursive relations (3.72) and (3.73), we solve for R_{k-1} and Q_{k-1} to get

$$R_{k-1}(z^{-1}) = \frac{\tau_k \tilde{Q}_k(z^{-1}) - R_k(z^{-1})}{\kappa_k^2}, \qquad (3.76)$$

$$Q_{k-1}(z^{-1}) = \frac{Q_k(z^{-1}) - \tau_k \tilde{R}_k(z^{-1})}{\kappa_k^2}. \qquad (3.77)$$

Since R_k and Q_k are polynomials of degree k while R_{k-1} and Q_{k-1} have degree k–1, the coefficient of the z^{-k} term on the right-hand side of Equations 3.76 and 3.77 must be zero. By imposing this condition, we obtain the following expression for the transmission coefficient τ_k:

$$\tau_k = \frac{r_k^{(k)}}{q_0^{(k)}} = \frac{q_k^{(k)}}{r_0^{(k)}}, \quad 0 \le k \le N, \tag{3.78}$$

where $r_k^{(k)}$ and $q_k^{(k)}$ denote the coefficients of the kth-power terms of R_k and Q_k, respectively.

In the design of a serially coupled microring filter, we assume that the drop port and through port transfer functions of the target filter are given by $H_d(z^{-1}) = K_0/Q(z^{-1})$ and $H_t(z^{-1}) = R(z^{-1})/Q(z^{-1})$, where K_0 is a constant and $R(z^{-1})$ and $Q(z^{-1})$ are polynomials of degree N of the form

$$R(z^{-1}) = r_N z^{-N} + r_{N-1} z^{-(N-1)} + r_1 z^{-1} + r_0, \tag{3.79}$$

$$Q(z^{-1}) = z^{-N} + q_{N-1} z^{-(N-1)} + q_1 z^{-1} + q_0. \tag{3.80}$$

We begin by calculating the transmission coefficient $\tau_N = r_N/q_0 = 1/r_0$ of the coupling junction between microring N and the output waveguide. Knowledge of τ_N and κ_N allows us to compute R_{N-1} and Q_{N-1} using Equations 3.76 and 3.77. The transmission coefficient τ_{N-1} is next obtained using Equation 3.78 and the procedure is repeated until all the coupling coefficients are obtained.

The above method requires that the transfer functions of the target filter be specified in terms of the z^{-1} variable. Two important types of all-poles filters whose transfer functions can be readily obtained in the inverse z-domain are the Butterworth and Chebyshev filters. Specifically, Butterworth filters have all reflection zeros (roots of $R(z^{-1})$) located at $z_k = 1$. On the other hand, the reflection zeros of a Chebyshev filter of order N are located at (Orta et al. 1995)

$$z_k = -\exp\left\{2j\cos^{-1}\left[\sin\left(\frac{\Delta\omega_R}{4\Delta\omega_{FSR}}\right)\cos\frac{(2k-1)\pi}{2N}\right]\right\}, \quad (k = 1...N) \tag{3.81}$$

where $\Delta\omega_R$ and $\Delta\omega_{FSR}$ are the ripple bandwidth and the FSR, respectively. From the given zeros, we can construct the polynomial $R(z^{-1})$ as follows:

$$R(z^{-1}) = \prod_{k=1}^{N}(z^{-1} - z_k). \tag{3.82}$$

Using the Feldtkeller relation in Equation 3.29, we have

$$Q(z^{-1})\tilde{Q}(z^{-1}) = R(z^{-1})\tilde{R}(z^{-1}) + z^{-N}K_0^2, \tag{3.83}$$

where the constant K_0 can be found from the 3 dB bandwidth $\Delta\omega_B$ of the filter. Specifically, writing $z_B^{-1} = \exp(-j\Delta\omega_B/2\Delta\omega_{FSR})$, we have

$$\left|H_d(z_B^{-1})\right|^2 = \frac{K_0^2}{K_0^2 + \left|R(z_B^{-1})\right|^2} = \frac{1}{2}, \tag{3.84}$$

from which we obtain $K_0 = |R(z_B^{-1})|$. To determine $Q(z^{-1})$, we find the roots of the polynomial on the right-hand side of Equation 3.83 and choose N roots with magnitude greater than unity[*] from which to construct the polynomial $Q(z^{-1})$.

As an example, we consider the design of a fifth-order microring Chebyshev filter with 3 dB bandwidth $\Delta\omega_B = 0.2\Delta\omega_{FSR}$ and ripple bandwidth $\Delta\omega_R = 0.1\Delta\omega_{FSR}$. We use Equation 3.81 to compute the zeros of the polynomial $R(z^{-1})$ and determine its coefficients to be $\{r_0, \ldots, r_5\} = \{-2.3874, 11.6448, -23.0047, 23.0047, -11.6448, 2.3874\}$. From the 3 dB bandwidth, we calculate $K_0 = 0.1507$ using Equation 3.84. Next we construct the polynomial $Q(z^{-1})$ using the Feldtkeller relation in Equation 3.83 and obtain the coefficients $\{q_0, \ldots, q_5\} = \{-5.6996, 18.2328, -24.8563, 17.7288, -6.5563, 1\}$. The filter is realized with an array of five serially coupled microring resonators with coupling coefficients $\{\kappa_0, \ldots, \kappa_5\} = \{0.908, 0.522, 0.343, 0.343, 0.522, 0.908\}$. The target spectral responses and those at the drop port and through port of the synthesized filter are shown in Figure 3.15.

---

[*] We choose roots lying outside the unit circle since we are operating in the inverse z-domain.

3.5 Parallel Cascaded ADMRs

In Section 3.1.2 we showed that an infinite array of cascaded ADMRs has transmission bands and stopbands that arise from both the microring and Bragg resonances. For an ADMR array of finite length, the transmission band has ripples due to impedance mismatch or reflections at the two terminations. By apodizing the coupling coefficients of the ADMRs, the ripples can be reduced or eliminated to produce a filter response with flat-top passband and smooth skirt roll-offs (Little et al. 2000). Apodization is a technique commonly employed in the design of Bragg grating filters (Hill and Meltz 1997). For a microring array, we can apodize either the input coupling coefficients (i.e., the coupling coefficients between the microrings and the input waveguide) or the output coupling coefficients, or both. Apodizing both coupling coefficients results in symmetrically coupled ADMRs which are simpler to analyze. In this section we use the transfer matrix method to derive the trans- fer functions of a finite array of cascaded ADMRs with symmetric couplings (Grover et al. 2002).

Figure 3.16 shows a schematic of an apodized array of N cascaded ADMRs. Each microring k in the array is coupled to the input and output bus waveguides with equal input and output field coupling coefficients κ_k. For simplicity, we also assume that all the microrings are identical with round-trip phase ϕ_{rt} and amplitude attenuation a_{rt}. Adjacent microrings are separated by two parallel waveguide sec- tions of length L, which is assumed to be greater than the microring

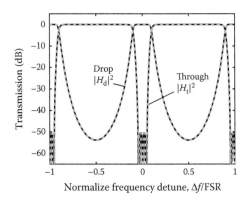

Figure 3.15 Spectral responses of a fifth-order Chebyshev serially cou- pled microring filter (black solid lines: target filter response; gray dashed lines: synthesized filter responses).

diameter so that there is no direct coupling between the two microrings. An input signal s_i applied to the input waveguide will be partially coupled to the microrings and the remaining power is transmitted to the through port. The light waves in the microrings are in turn coupled to the output waveguide where they combine to give the drop port signal s_d. Note that the signal that emerges at the drop port of each microring propagates backward to the previous microring, thereby forming an effective cavity of length L between two adjacent resonators. This feedback gives rise to additional resonances besides those of the microring resonators and, as a result, the transfer functions of the ADMR array will contain more poles than the number of microrings in the array.

With reference to Figure 3.16, we characterize each ADMR k in the array by a transfer matrix $\mathbf{M}_k$ defined as

$$\begin{bmatrix} c_k \\ d_k \end{bmatrix} = \mathbf{M}_k \begin{bmatrix} a_k \\ b_k \end{bmatrix}. \tag{3.85}$$

The elements of $\mathbf{M}_k$ are given by

$$M_{11}^{(k)} = \frac{\tau_k^2 - z^{-1}}{\tau_k(1 - z^{-1})}, \tag{3.86}$$

$$M_{21}^{(k)} = -M_{12}^{(k)} = \frac{-\kappa_k^2 z^{-1/2}}{\tau_k(1 - z^{-1})}, \tag{3.87}$$

$$M_{22}^{(k)} = \frac{1 - \tau_k^2 z^{-1}}{\tau_k(1 - z^{-1})}, \tag{3.88}$$

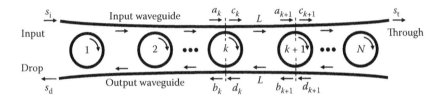

Figure 3.16 Schematic of an array of N cascaded add-drop microring resonators with coupling apodization.

where $z^{-1} = a_{rt}e^{-j\phi_{rt}}$ and $\phi_{rt} = \beta_r 2\pi R$ is the round-trip phase of the microring. The transfer matrix $\mathbf{P}$ of the two parallel waveguides of length L connecting two adjacent microrings is

$$\begin{bmatrix} a_{k+1} \\ b_{k+1} \end{bmatrix} = \begin{bmatrix} e^{-j\theta} & 0 \\ 0 & e^{j\theta} \end{bmatrix} \begin{bmatrix} c_k \\ d_k \end{bmatrix} \equiv \mathbf{P} \begin{bmatrix} c_k \\ d_k \end{bmatrix}, \tag{3.89}$$

where $\theta = \beta_b L$ is the phase shift and β_b is the propagation constant of the bus waveguides. For simplicity we have neglected loss in the bus waveguides. The total transfer matrix of the ADMR array is simply the product of the transfer matrices of the N stages,

$$\mathbf{T} = (\mathbf{M}_N \mathbf{P})(\mathbf{M}_{N-1}\mathbf{P})\cdots(\mathbf{M}_2 \mathbf{P})\mathbf{M}_1. \tag{3.90}$$

The transfer functions at the drop port and through port of the array are then obtained from $H_d = -T_{21}/T_{22}$ and $H_t = \det(\mathbf{T})/T_{22}$, respectively.

Analytical expressions for the transfer functions of a cascaded ADMR array can be obtained for certain values of the spacing L. For the special case where $L = \pi R + \lambda_0/4n_b$ (Little et al. 2000), we can write $\theta \approx \phi_{rt}/2 + \pi/2$, where we have assumed that the propagation constants of the bus waveguides and microring waveguides are equal ($\beta_b \approx \beta_r$), and that the phase term $\beta_r(\lambda_0/4n_b) \approx \pi/2$ changes much more slowly over one FSR compared to the round-trip phase ϕ_{rt} of the microring. With these approximations, we have $e^{-j\theta} = -jz^{-1/2}$, so the matrix $\mathbf{P}$ in Equation 3.89 becomes

$$\mathbf{P} = \begin{bmatrix} -jz^{-1/2} & 0 \\ 0 & jz^{1/2} \end{bmatrix} = jz^{1/2} \begin{bmatrix} -z^{-1} & 0 \\ 0 & 1 \end{bmatrix}. \tag{3.91}$$

If we define the transfer matrix of the first k microring stages as

$$\mathbf{T}_k = (\mathbf{M}_k \mathbf{P})(\mathbf{M}_{k-1}\mathbf{P})\cdots(\mathbf{M}_2 \mathbf{P})\mathbf{M}_1, \tag{3.92}$$

then starting from $\mathbf{T}_1 = \mathbf{M}_1$, we can use the recursive relation $\mathbf{T}_k = (\mathbf{M}_k \mathbf{P})\mathbf{T}_{k-1}$ to show that the transfer matrix $\mathbf{T}_N = \mathbf{T}$ of the entire array has the general form

$$\mathbf{T} = \frac{(jz^{1/2})^{N-1}}{\tau_1 \tau_2 \cdots \tau_N (1 - z^{-1})^N} \begin{bmatrix} z^{-1}R_N(z^{-1}) & -z^{-1/2}\tilde{P}_N(z^{-1}) \\ z^{-1/2}P_N(z^{-1}) & -\tilde{R}_N(z^{-1}) \end{bmatrix}, \tag{3.93}$$

and the polynomials in the matrix can be computed from the recursive relations

$$R_k = -(\tau_k^2 - z^{-1})z^{-1}R_{k-1} - \kappa_k^2 P_{k-1}, \tag{3.94}$$

$$\tilde{R}_k = (1 - \tau_k^2 z^{-1})\tilde{R}_{k-1} - \kappa_k^2 \tilde{P}_{k-1} z^{-2}, \tag{3.95}$$

$$P_k = -\kappa_k^2 z^{-2}R_{k-1} + (1 - \tau_k^2 z^{-1})P_{k-1}, \tag{3.96}$$

$$\tilde{P}_k = -\kappa_k^2 \tilde{R}_{k-1} - (\tau_k^2 - z^{-1})z^{-1}\tilde{P}_{k-1}. \tag{3.97}$$

The transfer functions at the drop port and through port of the cascaded ADMR array are obtained from the matrix **T** as follows:

$$H_d(z^{-1}) = -\frac{T_{21}}{T_{22}} = -\frac{z^{-1/2}P_N}{\tilde{R}_N}, \tag{3.98}$$

$$H_t(z^{-1}) = \frac{\det(\mathbf{T})}{T_{22}} = \tau_1\tau_2\cdots\tau_N(jz)^{-(N-1)/2}\frac{(1-z^{-1})^N}{\tilde{R}_N}. \tag{3.99}$$

Equation 3.99 indicates that the through port transfer function has N identical zeros at $z^{-1} = 1$, implying that the drop port response will have a maximally flat passband. On the other hand, since P_N and $\tilde{R}_N$ are polynomials of degree $2(N-1)$, the drop port transfer function in Equation 3.98 has $2(N-1)$ zeros and $2(N-1)$ poles. Since there are more poles and zeros in the transfer function than the number of design parameters (which are the N coupling coefficients), the poles and transmission zeros of the filter cannot be independently chosen.

In Figure 3.17a we plot the spectral responses at the drop port and through port of an unapodized array of $N = 7$ cascaded ADMRs with identical coupling coefficients $\kappa_k = 0.1$. The waveguide length L is chosen to be $L = \pi R + \lambda_0/4n_b$. We observe that a flat-top passband occurs at the resonant frequency of the microrings, but there are large side lobes in the drop port response. To suppress the side lobes, we apply apodization to the coupling coefficients in the array according to the Gaussian function $\kappa_k = 0.1\exp[-0.25(k-4)^2]$ (Little et al. 2000). The spectral responses of the apodized ADMR array are plotted in Figure 3.17b, which shows that the side lobes have been suppressed to produce a flat-top filter response with a smooth skirt roll-off.

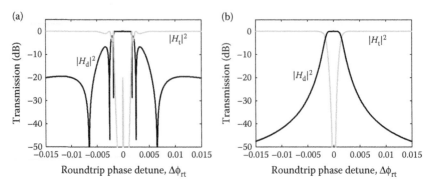

Figure 3.17   (a) Drop port and through port spectral responses of an array of seven cascaded ADMRs (a) with no coupling apodization ($\kappa_k = 0.1$) and (b) with Gaussian apodization of the coupling coefficients.

3.6   Parallel Cascaded Microring Doublets

In the previous section we saw that in an array of cascaded ADMRs, the drop port and through port signals of the microrings propagate in opposite directions in the two bus waveguides. This forms feedback loops in the array which give rise to more poles (or resonances) in the transfer functions than the number of resonators in the structure. Since the poles and zeros cannot be independently placed, it is generally not possible to exactly synthesize a given filter transfer function using cascaded ADMRs.

Arrays of cascaded ADMRs are examples of a feedback microring array, which consists of a parallel cascade of microring networks, typically of odd orders, whose output signals at the drop port and through port travel in opposite directions in the two connecting bus waveguides. Feedback microring arrays share the common feature that their transfer functions have more poles than the number of resonators in the structure. On the other hand, a feedforward microring array is a cascade of microring networks in which the output signals at the drop port and through port of each microring stage propagate in the same direction in the two connecting bus waveguides (Prabhu et al. 2008b). Since there is no feedback in the structure, the transfer functions of a feedforward array contain exactly the same number of poles as the number of microring resonators. These structures are more useful for filter applications than feedback arrays since they can be used to exactly realize a given transfer function.

The simplest example of a feedforward microring array is a parallel cascaded array of symmetric microring doublets, as shown in

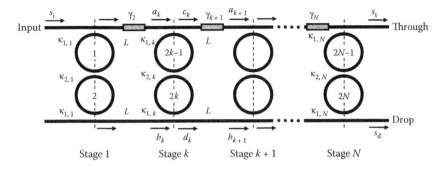

Figure 3.18 Schematic of a microring ladder filter consisting of N cascaded microring doublets.

Figure 3.18 (Liew and Van 2008). We refer to this structure simply as a microring ladder filter. As we will show below, the transfer function of a microring ladder filter contains transmission zeros that make them useful for realizing new classes of filters, such as inverse Chebyshev, pseudo-elliptic,[*] and linear phase filters, that are not possible with conventional CROW filters.

We will use the power coupling formalism and the transfer matrix method to analyze the microring ladder filter in Figure 3.18. The array consists of N microring doublet stages, for a total of $2N$ microring resonators. We assume the microrings to be lossless and synchronously tuned to the same resonant frequency. Each microring doublet in stage k is composed of two serially coupled microring resonators with symmetric bus-to-ring coupling coefficients $\kappa_{1,k}$ and ring-to-ring coupling coefficient $\kappa_{2,k}$. Connecting adjacent microring stages are two parallel bus waveguides of length L, with a possible π-phase shift in the upper waveguide with respect to the lower waveguide. We denote this π-phase shift by the factor $\gamma_k = e^{-j\pi} = -1$. If there is no phase shift, then $\gamma_k = 1$. The differential π-phase shifts are necessary for achieving destructive interference between the signal pathways in the upper and lower bus waveguides, thereby permitting transmission zeros to be realized in the drop port response of the filter.

---

[*] An inverse Chebyshev filter is a filter with equi-ripples in the stopband and no ripples in the passband. An elliptic or pseudo-elliptic filter is a filter with equi-ripples in both the passband and the stopband. The transfer function of an elliptic filter of order N has N poles and N transmission zeros, while a pseudo-elliptic filter has only up to $(N-2)$ transmission zeros.

Following the analysis of cascaded ADMRs in Section 3.5, we define the transfer matrix of the first k stages of the microring ladder filter as

$$\mathbf{T}_k = (\mathbf{M}_k \mathbf{P}_k)(\mathbf{M}_{k-1} \mathbf{P}_{k-1}) \cdots (\mathbf{M}_2 \mathbf{P}_2) \mathbf{M}_1, \tag{3.100}$$

where $\mathbf{M}_k$ denotes the transfer matrix of the kth microring doublet and $\mathbf{P}_k$ is the transfer matrix of the two connecting waveguides. In terms of the round-trip delay variable, the transfer matrix $\mathbf{M}_k$ of the symmetric microring doublet in stage k can be expressed in the form

$$\mathbf{M}_k = \frac{1}{G_k(z^{-1})} \begin{bmatrix} F_k(z^{-1}) & jK_k z^{-1} \\ jK_k z^{-1} & F_k(z^{-1}) \end{bmatrix}, \tag{3.101}$$

where $z^{-1} = a_{rt} e^{-j\phi_{rt}}$ and

$$K_k = \kappa_{1,k}^2 \kappa_{2,k}, \tag{3.102}$$

$$F_k(z^{-1}) = \tau_{1,k} - \tau_{2,k}(1 + \tau_{1,k}^2)z^{-1} + \tau_{1,k} z^{-2}, \tag{3.103}$$

$$G_k(z^{-1}) = 1 - 2\tau_{1,k}\tau_{2,k} z^{-1} + \tau_{1,k}^2 z^{-2}. \tag{3.104}$$

In the above expressions, $\tau_{1,k} = \sqrt{1 - \kappa_{1,k}^2}$ and $\tau_{2,k} = \sqrt{1 - \kappa_{2,k}^2}$ are the transmission coefficients of the respective coupling junctions. The transfer matrix $\mathbf{P}_k$ of the two bus waveguides of length L connecting microring stages k and $k+1$ with differential phase shift factor γ_k is given by

$$\mathbf{P}_k = e^{-j\beta_b L} \begin{bmatrix} \gamma_k & 0 \\ 0 & 1 \end{bmatrix}, \tag{3.105}$$

where β_b is the propagation constant of the waveguides. Note that if there is no differential phase shift ($\gamma_k = 1$), then $\mathbf{P}_k = e^{-j\beta_b L}\mathbf{I}$, where $\mathbf{I}$ is the identity matrix. Starting from $\mathbf{T}_1 = \mathbf{M}_1$ and using the recursive relation $\mathbf{T}_k = (\mathbf{M}_k \mathbf{P}_k)\mathbf{T}_{k-1}$, we can show that the transfer matrix $\mathbf{T}_N = \mathbf{T}$ of N cascaded stages can be expressed in the form

$$\mathbf{T}_N = \frac{e^{-j\beta_b(N-1)L}}{Q_N(z^{-1})} \begin{bmatrix} R_N(z^{-1}) & j\sigma_N P_N(z^{-1})z^{-1} \\ jP_N(z^{-1})z^{-1} & \sigma_N R_N(z^{-1}) \end{bmatrix}, \tag{3.106}$$

where $\sigma_N = \gamma_2 \gamma_3 \ldots \gamma_N = \pm 1$, and P_N, R_N, and Q_N are polynomials satisfying the recursive relations

$$P_k = \gamma_k K_k R_{k-1} + F_k P_{k-1}, \tag{3.107}$$

$$R_k = \gamma_k F_k R_{k-1} - K_k P_{k-1} z^{-2}, \tag{3.108}$$

$$Q_k = G_k Q_{k-1} = \prod_{n=1}^{k} G_n. \tag{3.109}$$

The transfer functions at the drop port and through port of the microring ladder filter can be obtained from the transfer matrix **T**. Neglecting the common phase factor $e^{-j\beta_b(N-1)L}$ in Equation 3.106, we have

$$H_d(z^{-1}) = T_{21} = \frac{jP_N(z^{-1})z^{-1}}{Q_N(z^{-1})}, \tag{3.110}$$

$$H_t(z^{-1}) = T_{11} = \frac{R_N(z^{-1})}{Q_N(z^{-1})}. \tag{3.111}$$

Starting with the first microring stage with $P_1 = K_1$, $R_1 = F_1$, and $Q_1 = G_1$, we can deduce from the recursive relations in Equations 3.107 through 3.109 that P_N is a polynomial of degree $2(N-1)$, while R_N and Q_N are polynomials of degree $2N$. Furthermore, since K_k and F_k are even-degree and self-para-conjugate polynomials,[*] both P_N and R_N are also of even-degree and self-para-conjugates. This implies that the roots of P_N and R_N appear in both complex conjugate pairs and para-conjugate pairs, for example, as $\{z_k, 1/z_k^*\}$ and $\{z_k^*, 1/z_k\}$.[†] Note that a pair of conjugate roots located on the unit circle also satisfies this property. Thus a microring ladder filter with N stages can realize a drop port transfer function with $2N$ poles and up to $2(N-1)$ transmission zeros that appear in complex and para-conjugate pairs. Each microring doublet in the array is responsible for generating a pair of complex conjugate poles in the transfer function.

Microring ladder filters can be synthesized using a procedure based on the order reduction technique similar to the synthesis

---

[*] A polynomial $P(z^{-1})$ of degree N is self-para-conjugate if $P(z^{-1}) = \pm\tilde{P}(z^{-1})$. The polynomial is even self-para-conjugate if the coefficients of the kth and $(N-k)$th power terms are equal; it is odd self-para-conjugate if these coefficients are equal in magnitude but have opposite signs.

[†] In the s-domain, the transfer functions at the drop port and through port of the microring ladder filter have transmission zeros that are quadrantally symmetric, i.e., they appear as $\{\pm z, \pm z^*\}$ (Liew and Van 2008).

method developed for serially coupled microring filters in Section 3.4.4 (Liew and Van 2008). Suppose the target filter responses are described by the transfer functions H_d and H_t of the form in Equations (3.110) and (3.111), where P_N is a polynomial of degree $2(N-1)$, and R_N and Q_N are polynomials of degree $2N$. The roots of P_N and R_N appear in conjugate pairs that either lie on the unit circle or are also paired by their para-conjugates. Many common filters with symmetric spectral responses, such as pseudo-elliptic and inverse Chebyshev filters, satisfy the first property, that is, their transmission and reflection responses have transmission nulls that are symmetrically located about the center frequency. The second property (para-conjugate zeros) is typically satisfied by filters with linear phase response.

The transfer functions H_d and H_t specified above can be realized by an array of N microring doublets. The synthesis procedure begins with the determination of the coupling parameters of the last microring stage N, and proceeds backward until the first stage is reached. Each microring doublet in stage k is completely characterized by a pair of conjugate poles $\{p_k, p_k^*\}$ in the transfer functions. Specifically, for each stage k, we select a pair of roots $\{p_k, p_k^*\}$ from the polynomial Q_k to be the poles of the microring doublet. Since these poles are the roots of the polynomial G_k in Equation 3.104, they are given by

$$\{p_k, p_k^*\} = (\tau_{2,k} \pm j\kappa_{2,k})/\tau_{1,k}. \tag{3.112}$$

From the above expression, we can solve for the transmission coefficients $\tau_{1,k}$ and $\tau_{2,k}$ of the microring doublet to get $\tau_{1,k} = |p_k|^{-2}$ and $\tau_{2,k} = \tau_{1,k}\text{Re}\{p_k\}$. Note that since the poles of a filter are typically located in the right half of the (inverse) z-plane near the real axis, the transmission coefficients $\tau_{1,k}$ and $\tau_{2,k}$ obtained from these formulas are always positive. Knowledge of $\tau_{1,k}$ and $\tau_{2,k}$ allows us to construct the transfer matrix $\mathbf{M}_k$ of stage k as given by Equation 3.101, which is then de-embedded from the array. The transfer matrix $\mathbf{T}_{k-1}$ of the remaining $k-1$ stages is obtained by solving for the polynomials P_{k-1} and R_{k-1} from Equations 3.107 and 3.108. The results are

$$P_{k-1} = \frac{F_k P_k - K_k R_k}{G_k \tilde{G}_k}, \tag{3.113}$$

$$R_{k-1} = \frac{F_k R_k + K_k P_k z^{-2}}{\gamma_k G_k \tilde{G}_k}, \tag{3.114}$$

where $\tilde{G}_k = \tau_{1,k}^2 - 2\tau_{1,k}\tau_{2,k}z^{-1} + z^{-2}$ is the Hermitian conjugate of G_k. The differential phase factor γ_k of stage k is chosen to be either 1 or -1 such that $R_{k-1}(p_{k-1})/P_{k-1}(p_{k-1}) = j$, where $\{p_{k-1}, p_{k-1}^*\}$ is the pair of poles selected for the next stage, $k-1$. The procedure is then repeated with the parameter extraction of stage $k-1$ until the first stage is reached.

Since the transfer matrices $\mathbf{M}_k$ in Equation 3.101 are symmetric with equal diagonal elements, they commute with each other, $\mathbf{M}_k\mathbf{M}_{k-1} = \mathbf{M}_{k-1}\mathbf{M}_k$. Thus, two microring doublets connected by two parallel waveguides with no differential phase shift can be interchanged without affecting the response of the filter. On the other hand, if two microring doublets are connected by two waveguides with a differential π-phase shift, then it can be shown that $\mathbf{M}_k\mathbf{P}\mathbf{M}_{k-1} = \mathbf{P}(\mathbf{M}_{k-1}\mathbf{P}\mathbf{M}_k)\mathbf{P}$, where $\mathbf{P} = \mathrm{diag}[-1, 1]$. In this case the order of the two stages can be exchanged if a π-phase shift is also added before and after the two stages. These commutative properties of the matrices $\mathbf{M}_k$ imply that the order in which the roots of Q_N are used to determine the coupling parameters of the microring stages does not affect the final filter design, except for a possible permutation of the stage order and a different distribution of the phase shift elements in the array. Moreover, it can also be shown that by applying the above commutative properties, one can always reduce the number of π-phase shifts in a given microring ladder filter to at most one. A procedure for performing this phase shift reduction is given in Prabhu et al. (2008b).

As an example, we consider the design of an eighth-order pseudo-elliptic filter with a 100 GHz bandwidth, 0.5 dB in-band ripple and -60 dB out-of-band rejection. Using a suitable filter approximation technique (e.g., Martinez and Parks 1978), we obtain a transfer function with eight poles and six zeros of the form given by Equation 3.110 which satisfy these specifications. The roots of the polynomials P_N and Q_N are listed in Table 3.1. Using the Feldtkeller relation in Equation 3.29, we also obtain the transfer function for the reflection response of the form in Equation 3.111. The roots of the polynomial R_N are also given in Table 3.1. A plot of the pole-zero diagram of the filter in the inverse z-plane is shown in Figure 3.19a and the target spectral responses are shown in Figure 3.19b. We synthesize the transfer functions using a cascaded array of $N = 4$ microring doublets. Assuming that the microring resonators have an FSR of 0.5 THz, we compute the transmission coefficients for each microring

Table 3.1   Poles and Zeros of an Eighth-Order
Pseudo-Elliptic Filter

Transmission Zeros (Roots of P_N)	Reflection Zeros (Roots of R_N)	Poles (Roots of Q_N)
$0.3435 \pm j0.9391$	$0.8129 \pm j0.5825$	$0.7855 \pm j0.5851$
$0.7238 \pm j0.6901$	$0.8466 \pm j0.5323$	$0.7691 \pm j0.5161$
$0.6593 \pm j0.7519$	$0.9159 \pm j0.4013$	$0.7609 \pm j0.3709$
	$0.9879 \pm j0.1554$	$0.7608 \pm j0.1370$

doublet and list their values in Table 3.2 along with the phase shift factors. The factor $\gamma_k = -1$ in stage 3 indicates that there is a π-phase shift between microring doublets 2 and 3. The spectral responses of the microring ladder filter are shown in Figure 3.19b, which shows good agreement between the target and synthesized responses.

Experimentally, a fourth-order microring ladder filter consisting of two cascaded microring doublets has been demonstrated in the SOI material (Masilamani and Van 2012). The filter was designed to realize a fourth-order pseudo-elliptic transfer function with two transmission zeros. Although the device exhibited a fourth-order skirt roll-off, fabrication errors caused the zeros to be displaced from their designed locations so that the measured filter response did not show deep transmission nulls associated with the zeros. Nevertheless, the microring ladder configuration holds promising potential for realizing high-order optical filters with sharp band transitions because they can be constructed and optimized stage by stage, with each stage consisting of a simple microring doublet. Another important advantage of the ladder configuration is that it does not require negative coupling coefficients to realize transfer functions with transmission zeros located on the unit circle (or on the $j\omega$-axis in the s-domain), as in the example above. This is in

Table 3.2   Transmission Coefficients of the
Microring Ladder Filter with Four Stages

	$\tau_{1,k}$	$\tau_{2,k}$	γ_k
Stage 1	0.7731	0.9842	
Stage 2	0.9263	0.8304	1
Stage 3	0.8464	0.8989	−1
Stage 4	0.9795	0.8019	1

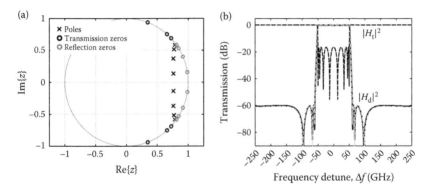

Figure 3.19   (a) Pole-zero diagram of an eighth-order pseudo-elliptic filter and (b) target filter responses (gray dashed lines) and spectral responses of the synthesized microring ladder filter consisting of four cascaded microring doublets (black solid lines).

contrast to filters based on 2D microring coupling topologies, which are discussed in the next section.

An alternative filter structure that is closely related to the microring ladder filter is a Mach–Zehnder interferometer (MZI) whose arms are loaded with APMRs (Madsen 1998). These microring-loaded MZI structures (or RMZIs) are the photonic realizations of a general class of digital filters known as sum–difference all-pass filters, which can realize Nth-order transfer functions containing up to N–2 transmission zeros. In fact, it can be shown that an array of N cascaded microring doublets is equivalent to an MZI with each arm loaded with N APMRs (Van 2009). A significant difference between these two structures is that the microring resonators in the RMZI filter generally have non-zero phase detunes, whereas they are synchronously tuned in the microring ladder filter.

3.7   2D Networks of CMRs

CMR networks of 2D coupling topology are extensions of serially coupled microring filters, which can be regarded as microring networks of 1D coupling topology. In Section 3.4, we showed that the drop port transfer function of a serially coupled microring filter is restricted to those with all poles and no transmission zeros. This is due to the fact that near a resonant frequency, all signal pathways through the 1D microring array are nearly in phase, so it is not possible to achieve destructive interference of the signals at the drop port to produce transmission nulls. In contrast, in a 2D CMR network, light follows different pathways through the device which can interfere destructively with each other at the output to produce transmission nulls at certain frequencies. As a result, 2D microring coupling topologies allow more complex transfer functions to be realized than possible with the 1D topology. Examples of filters that can be synthesized with 2D CMR networks include inverse Chebyshev and pseudo-elliptic filters with sharp band transitions, and linear phase filters with constant group-delay response.

In this section we will first develop techniques for analyzing and designing 2D CMR filters based on the energy coupling formalism. This will be followed by an analysis of broadband or strongly coupled microring networks in terms of power coupling. Important differences in the spectral characteristics of the device obtained from the two methods will also be highlighted.

3.7.1   Energy coupling analysis of 2D CMR networks

We consider a general network of N direct-coupled microring resonators arranged in a 2D square lattice as shown in Figure 3.20. Other 2D arrangements of microrings are also possible but we will focus on the 2D square lattice because it is the simplest coupling topology to implement in practice. In general, there is no restriction on the coupling topology except that it should not give rise to coupling between counter-propagating modes in the microrings. This precludes device configurations that contain coupling loops with an

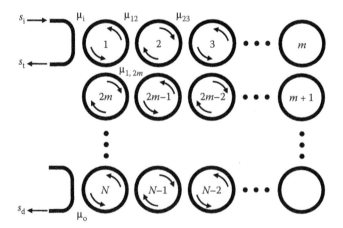

Figure 3.20   Schematic of a network of N microring resonators in a 2D square lattice coupling topology.

odd number of microrings, such as the triplet shown in Figure 3.21b. In such a loop configuration, coupling between counter-propagating modes gives rise to a back-reflected wave at the input port of the device, which is undesirable for most integrated optics applications.

We label the microrings in the network from 1 to N as shown in Figure 3.20, although the order of numbering does not affect the analysis that follows. For simplicity we assume that the microrings have the same intrinsic decay rate, $\gamma = \alpha v_g/2$, due to propagation loss α in the waveguides. However, we allow the microrings to have non-identical resonant frequencies ω_i to account for possible resonance mismatches among them. Coupling between two adjacent microrings i and j is denoted by the energy coupling coefficient $\mu_{i,j}$. Microrings 1 and N are also coupled to the input and output waveguides via input and output coupling coefficients μ_i and μ_o, respectively.

We denote the wave amplitude in each microring i as a_i, which is normalized so that $|a_i|^2$ gives the energy stored in the resonator. The signals s_i, s_t, and s_d in the input and output waveguides denote the input, through, and drop signals, respectively. These signals represent the rates of energy supplied to or drawn from the network and are thus normalized with respect to power. By examining the coupled mode equations in Equation 3.38 for a serially coupled microring filter, we can generalize them for a 2D CMR network as follows (Van 2007):

$$\frac{d}{dt}\begin{bmatrix} a_1 \\ a_2 \\ a_3 \\ \vdots \\ a_N \end{bmatrix} = \begin{bmatrix} (j\omega_1 - \gamma_1) & -j\mu_{1,2} & -j\mu_{1,3} & \cdots & -j\mu_{1,N} \\ -j\mu_{1,2} & (j\omega_2 - \gamma) & -j\mu_{2,3} & \cdots & -j\mu_{2,N} \\ -j\mu_{1,3} & -j\mu_{2,3} & (j\omega_3 - \gamma) & \cdots & -j\mu_{3,N} \\ \vdots & \vdots & \vdots & \ddots & \vdots \\ -j\mu_{1,N} & -j\mu_{2,N} & -j\mu_{3,N} & \cdots & (j\omega_N - \gamma_N) \end{bmatrix}\begin{bmatrix} a_1 \\ a_2 \\ a_3 \\ \vdots \\ a_N \end{bmatrix} + \begin{bmatrix} -j\mu_i s_i \\ 0 \\ 0 \\ \vdots \\ 0 \end{bmatrix}.$$

$$(3.115)$$

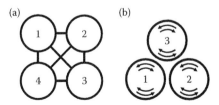

Figure 3.21 Examples of unrealizable or undesirable microring coupling topologies: (a) a quadruplet with cross couplings and (b) a triplet that causes coupling between counter-propagating modes in the microrings.

The above equation assumes the most general coupling topology where each microring i is coupled to every other microring j in the network via coupling coefficient $\mu_{i,j}$. If there is no coupling between microrings i and j, then $\mu_{i,j} = 0$. The amplitude decay rates γ_1 and γ_N represent the total decay rates in resonators 1 and N, respectively, due to both intrinsic loss and coupling to the input or output waveguide,

$$\gamma_{(1,N)} = \gamma + \gamma_{(i,o)}, \tag{3.116}$$

where $\gamma_{(i,o)} = \mu_{(i,o)}^2/2$.

To obtain the frequency response of the device, we assume a harmonic input excitation of the form $s_i \sim e^{j\omega t}$. The signals in the microrings will also have the same time dependence, $a_i \sim e^{j\omega t}$. The input frequency ω is detuned from the resonant frequency of microring i by an amount given by

$$\omega - \omega_i = (\omega - \omega_0) - (\omega_i - \omega_0) = \Delta\omega - \delta\omega_i, \tag{3.117}$$

where $\delta\omega_i$ is the detune of the resonant frequency of microring i from a center frequency ω_0 (typically chosen as the center frequency of the filter passband). Substituting the above expressions into Equation 3.115, we can write the resulting equation in matrix form as

$$(s\mathbf{I} + \mathbf{L} + j\mathbf{M})a = s, \tag{3.118}$$

where $s = j\Delta\omega + \gamma$, $a = [a_1, a_2 \ldots a_{N-1}, a_N]^T$, $s = [-j\mu_i s_i, 0, \ldots 0, 0]^T$, and $\mathbf{I}$ is the $N \times N$ identity matrix. The matrix $\mathbf{L}$ is a diagonal matrix that accounts for couplings to the input and output waveguides,

$$\mathbf{L} = \text{diag}[\gamma_i, \ 0, \ \ldots \ 0, \ \gamma_o], \tag{3.119}$$

and $\mathbf{M}$ is an $N \times N$ symmetric coupling matrix,

$$\mathbf{M} = \begin{bmatrix} \delta\omega_1 & \mu_{1,2} & \mu_{1,3} & \cdots & \mu_{1,N} \\ \mu_{1,2} & \delta\omega_2 & \mu_{2,3} & \cdots & \mu_{2,N} \\ \mu_{1,3} & \mu_{2,3} & \delta\omega_3 & \cdots & \mu_{3,N} \\ \vdots & \vdots & \vdots & \ddots & \vdots \\ \mu_{1,N} & \mu_{2,N} & \mu_{3,N} & \cdots & \delta\omega_N \end{bmatrix}. \tag{3.120}$$

For 1D CMR (or CROW) filters, the coupling matrix $\mathbf{M}$ has a simple tridiagonal form.

The solution of the matrix equation (3.118) can be expressed in closed form. We diagonalize the matrix $-(L + jM)$ in the form $-(L + jM) = Q \cdot D \cdot Q^{-1}$, where D is a diagonal matrix containing the eigenvalues of $-(L + jM)$ and Q is a matrix containing the corresponding eigenvectors. Making this substitution in Equation 3.118 and solving for a, we get

$$a = Q(sI - D)^{-1}Q^{-1}s. \tag{3.121}$$

The wave amplitude a_i can be explicitly expressed as

$$a_i = -j\mu_i s_i \sum_{k=1}^{N} \frac{Q_{i,k}Q_{k,1}^{-1}}{s - p_k}, \quad i = 1 \text{ to } N, \tag{3.122}$$

where p_k is the kth diagonal element of D, and $Q_{i,k}$ and $Q_{k,1}^{-1}$ are the matrix elements of Q and Q^{-1}, respectively. Using the relations $s_t = s_i - j\mu_i a_1$ and $s_d = -j\mu_o a_N$, we obtain the following expressions for the transfer functions at the through port and drop port of the CMR network:

$$H_t(s) = 1 - \mu_i^2 \sum_{k=1}^{N} \frac{Q_{1,k}Q_{k,1}^{-1}}{s - p_k} \equiv \frac{R(s)}{Q(s)}, \tag{3.123}$$

$$H_d(s) = -\mu_i\mu_o \sum_{k=1}^{N} \frac{Q_{N,k}Q_{k,1}^{-1}}{s - p_k} \equiv \frac{P(s)}{Q(s)}. \tag{3.124}$$

The above equations show that the 2D CMR filter has N poles which, in the absence of loss ($\gamma = 0$), are given by the eigenvalues p_k of the matrix $-(L + jM)$. The numerator polynomial $R(s)$ of the through port transfer function has degree N, while the numerator polynomial $P(s)$ of the drop port transfer function has a maximum degree of $N - 2$. The latter result follows from the fact that the coefficient of the highest power term of $P(s)$—that is, the $(N-1)$th power term—is zero,

$$\sum_{k=1}^{N} Q_{N,k}Q_{k,1}^{-1} = 0,$$

since it is the product of row N of Q and column 1 of Q^{-1}. Thus a 2D network of N microrings can realize an Nth-order filter transfer function containing up to $N - 2$ transmission zeros.

If we take the absolute square of each term in the summation in Equation 3.124, the result has the form of a Lorentzian resonance:

$$\left| \mu_i \mu_o \frac{Q_{N,k} Q_{k,1}^{-1}}{j(\omega - \omega_0) + \gamma - p_k} \right|^2 \propto \frac{1}{(\omega - \Omega_k)^2 + \Gamma_k^2}, \tag{3.125}$$

where $\Omega_k = \omega_0 - \text{Im}\{p_k\}$ and $\Gamma_k = \gamma + \text{Re}\{p_k\}$. Thus the spectral response of a 2D CMR network can be regarded as consisting of a sum of Lorentzian resonances centered at frequencies Ω_k with linewidths Γ_k.[*] The Lorentzian response is characteristic of the weak coupling approximation assumed in the energy coupling formalism.

3.7.2   Energy coupling synthesis of 2D CMR networks

We consider next the problem of synthesizing a 2D CMR network to achieve a target filter response. The synthesis problem was first considered in Van (2007) in which the CMR network was modeled by an equivalent electrical circuit. By applying a suitable electrical filter synthesis method (such as in Atia et al. 1974), the microring coupling parameters could be determined for a given transfer function. In this section, we will develop a simpler and more direct method for synthesizing 2D CMR networks without making use of an equivalent electrical circuit.

We consider a lossless[†] 2D CMR filter consisting of N microring resonators with identical resonant frequency ω_0. Setting $\gamma = 0$ in Equation 3.115, we rewrite the coupled mode equations for the CMR in the form

$$\frac{d}{dt}\begin{bmatrix} a_1 \\ a_2 \\ a_3 \\ \vdots \\ a_N \end{bmatrix} = \begin{bmatrix} j\omega_0 & -j\mu_{1,2} & -j\mu_{1,3} & \cdots & -j\mu_{1,N} \\ -j\mu_{1,2} & j\omega_0 & -j\mu_{2,3} & \cdots & -j\mu_{2,N} \\ -j\mu_{1,3} & -j\mu_{2,3} & j\omega_0 & \cdots & -j\mu_{3,N} \\ \vdots & \vdots & \vdots & \ddots & \vdots \\ -j\mu_{1,N} & -j\mu_{2,N} & -j\mu_{3,N} & \cdots & j\omega_0 \end{bmatrix}\begin{bmatrix} a_1 \\ a_2 \\ a_3 \\ \vdots \\ a_N \end{bmatrix} + \begin{bmatrix} u_1 \\ 0 \\ 0 \\ \vdots \\ u_N \end{bmatrix}, \tag{3.126}$$

---

[*] This statement is accurate if the resonances are far apart. When the resonances are close together or overlap, coupling between them gives rise to a spectral response that drastically deviates from the Lorentzian shape.

[†] For CMR filters with uniform microring loss, the transfer functions can be predistorted to compensate for the effect of loss before applying the synthesis procedure. The predistortion technique for synthesizing microring filters with loss can be found in Prabhu and Van (2008).

where

$$u_1 = -\gamma_i a_1 - j\mu_i s_i = -j\mu_i\left(s_i - \frac{j\mu_i}{2}a_1\right), \tag{3.127}$$

$$u_N = -\gamma_o a_N - j\mu_o s_a = -j\mu_o\left(s_a - \frac{j\mu_o}{2}a_N\right). \tag{3.128}$$

In Equation 3.128 s_a is the signal applied to the add port of the filter. The quantities $|u_1|^2$ and $|u_N|^2$ represent the net powers supplied to microrings 1 and N, respectively, from the input and output waveguides. Assuming signals with harmonic time dependence $e^{j\omega t}$, we can express Equation 3.126 as

$$(s\mathbf{I} + j\mathbf{M})a = u, \tag{3.129}$$

where $s = j(\omega - \omega_0)$, $u = [u_1, 0, ...0, u_N]^T$, and $\mathbf{M}$ is the coupling matrix in Equation 3.120 with zero frequency detunes,

$$\mathbf{M} = \begin{bmatrix} 0 & \mu_{1,2} & \mu_{1,3} & \cdots & \mu_{1,N} \\ \mu_{1,2} & 0 & \mu_{2,3} & \cdots & \mu_{2,N} \\ \mu_{1,3} & \mu_{2,3} & 0 & \cdots & \mu_{3,N} \\ \vdots & \vdots & \vdots & \ddots & \vdots \\ \mu_{1,N} & \mu_{2,N} & \mu_{3,N} & \cdots & 0 \end{bmatrix}. \tag{3.130}$$

Since $\mathbf{M}$ is real and symmetric, we can diagonalize it in the form $\mathbf{M} = \mathbf{W} \cdot \mathbf{\Lambda} \cdot \mathbf{W}^T$, where $\mathbf{\Lambda}$ is a diagonal matrix containing the eigenvalues and $\mathbf{W}$ is a real unitary matrix containing the eigenvectors. Making this substitution in Equation 3.129 and solving for a, we get

$$a = \mathbf{W}(s\mathbf{I} + j\mathbf{\Lambda})^{-1}\mathbf{W}^T u. \tag{3.131}$$

Specifically, a_1 and a_N are given by

$$a_1 = u_1 \sum_{k=1}^{N} \frac{W_{1,k}^2}{s + j\lambda_k} + u_N \sum_{k=1}^{N} \frac{W_{1,k}W_{N,k}}{s + j\lambda_k} \equiv A_{11}u_1 + A_{1N}u_N, \tag{3.132}$$

$$a_N = u_1 \sum_{k=1}^{N} \frac{W_{N,k}W_{1,k}}{s + j\lambda_k} + u_N \sum_{k=1}^{N} \frac{W_{N,k}^2}{s + j\lambda_k} \equiv A_{N1}u_1 + A_{NN}u_N, \tag{3.133}$$

where λk are the eigenvalues of $\mathbf{M}$ and $W_{i,j}$ are the elements of the matrix $\mathbf{W}$. If we remove the output waveguide so that the CMR becomes an all-pass network, then $\mu_o = 0$, which also gives $u_N = 0$. Under this condition, we obtain from Equations 3.132 and 3.133 the following expressions for the network parameters A_{11} and A_{N1}:

$$A_{11} = \left.\frac{a_1}{u_1}\right|_{\mu_o=0} = \sum_{k=1}^{N} \frac{W_{1,k}^2}{s + j\lambda_k}, \tag{3.134}$$

$$A_{N1} = \left.\frac{a_N}{u_1}\right|_{\mu_o=0} = \sum_{k=1}^{N} \frac{W_{N,k}W_{1,k}}{s + j\lambda_k}. \tag{3.135}$$

Next we establish the relationships between the parameters A_{11} and A_{N1} and the specified drop port and through port transfer functions of the filter. The input and output signals of the CMR filter are related by a scattering matrix $\mathbf{S}$ defined by

$$\begin{bmatrix} s_t \\ s_d \end{bmatrix} = \begin{bmatrix} S_{11}(s) & S_{12}(s) \\ S_{21}(s) & S_{22}(s) \end{bmatrix} \begin{bmatrix} s_i \\ s_a \end{bmatrix}. \tag{3.136}$$

From Equations 3.127 and 3.128, we can express s_i and s_a in terms of u_1 and u_N as

$$s_i = -\frac{1}{j\mu_i}\left(u_1 + \frac{\mu_i^2}{2}a_1\right), \tag{3.137}$$

$$s_a = -\frac{1}{j\mu_o}\left(u_N + \frac{\mu_o^2}{2}a_N\right). \tag{3.138}$$

At the input and output bus coupling junctions we also have $s_t = s_i - j\mu_i a_1$ and $s_d = s_a - j\mu_o a_N$. Substituting the expressions for s_i and s_a in Equations 3.137 and 3.138 into these relations, we get

$$s_t = -\frac{1}{j\mu_i}\left(u_1 - \frac{\mu_i^2}{2}a_1\right), \tag{3.139}$$

$$s_d = -\frac{1}{j\mu_o}\left(u_N - \frac{\mu_o^2}{2}a_N\right). \tag{3.140}$$

Equations 3.137 through 3.140 link the input and output signals s_i, s_a, s_t, and s_d of the CMR network to the newly defined signals

u_1 and u_N. Substituting these expressions into Equation 3.136 and solving for a_1 and a_N, we obtain

$$a_1 = \frac{2}{\mu_i^2}\left(\frac{1-S_{11}+S_{22}-\Delta_S}{1+S_{11}+S_{22}+\Delta_S}\right)u_1 - \frac{4}{\mu_i\mu_o}\left(\frac{S_{12}}{1+S_{11}+S_{22}+\Delta_S}\right)u_N, \quad (3.141)$$

$$a_N = -\frac{4}{\mu_i\mu_o}\left(\frac{S_{21}}{1+S_{11}+S_{22}+\Delta_S}\right)u_1 + \frac{2}{\mu_o^2}\left(\frac{1+S_{11}-S_{22}-\Delta_S}{1+S_{11}+S_{22}+\Delta_S}\right)u_N,$$
$$(3.142)$$

where $\Delta_S = S_{11}S_{22} - S_{12}S_{21}$ is the determinant of the scattering matrix. Comparing the above equations to Equations 3.132 and 3.133, we get

$$A_{11} = \frac{a_1}{u_1}\bigg|_{u_N=0} = \frac{2}{\mu_i^2}\left(\frac{1-S_{11}+S_{22}-\Delta_S}{1+S_{11}+S_{22}+\Delta_S}\right), \quad (3.143)$$

$$A_{N1} = \frac{a_N}{u_1}\bigg|_{u_N=0} = -\frac{4}{\mu_i\mu_o}\left(\frac{S_{21}}{1+S_{11}+S_{22}+\Delta_S}\right). \quad (3.144)$$

The above equations allow us to compute network parameters A_{11} and A_{N1} from the scattering parameters of the CMR filter.

Since an Nth-order CMR filter is a linear passive network, its scattering matrix has the general form

$$\mathbf{S} = \frac{1}{Q(s)}\begin{bmatrix} R(s) & \sigma jP(-s) \\ jP(s) & \sigma R(-s) \end{bmatrix}, \quad (3.145)$$

where $\sigma = +1/-1$ for N even/odd. The polynomials R and Q have degree N, while P has maximum degree N–2. The determinant of **S** is given by

$$\Delta_S = \frac{\sigma[R(s)R(-s)+P(s)P(-s)]}{Q^2(s)} = \sigma\frac{Q(-s)}{Q(s)}, \quad (3.146)$$

where we have made use of the Feldtkeller relation in Equation 3.26. By writing the elements of the scattering matrix in Equations 3.143 and 3.144 in terms of the polynomials P, R, and Q, we obtain the following expressions for A_{11} and A_{N1}:

$$A_{11} = \frac{2}{\mu_i^2}\left[\frac{Q(s)-R(s)+\sigma R(-s)-\sigma Q(-s)}{Q(s)+R(s)+\sigma R(-s)+\sigma Q(-s)}\right] \equiv \frac{M(s)}{D(s)}, \quad (3.147)$$

$$A_{N1} = -\frac{4}{\mu_i \mu_o} \left[\frac{jP(s)}{Q(s) + R(s) + \sigma R(-s) + \sigma Q(-s)} \right] = -\frac{4}{\mu_i \mu_o} \frac{jP(s)}{D(s)}.$$

(3.148)

In the above expressions, the polynomials $M(s)$ and $D(s)$ have maximum degree N with the leading coefficient of $D(s)$ assumed to be 1. Equations 3.147 and 3.148 allow us to determine the parameters A_{11} and A_{N1} of the CMR network from the specified transfer functions of the filter.

In the synthesis procedure of a 2D CMR filter, given the drop port and through port transfer functions of the filter of the form

$$H_d(s) = \frac{jP(s)}{Q(s)} = \frac{j(p_{N-2}s^{N-2} + p_{N-3}s^{N-3} + \cdots + p_1 s + p_0)}{s^N + q_{N-1}s^{N-1} + \cdots + q_1 s + q_0},$$

(3.149)

$$H_t(s) = \frac{R(s)}{Q(s)} = \frac{r_N s^N + r_{N-1}s^{N-1} + \cdots + r_1 s + r_0}{s^N + q_{N-1}s^{N-1} + \cdots + q_1 s + q_0},$$

(3.150)

we can determine the coupling coefficients of the CMR network as follows. Since the eigenvector matrix $\mathbf{W}$ is a unitary matrix, its rows have unity magnitude. Specifically, for row 1 we have

$$\sum_{k=1}^{N} W_{1,k}^2 = 1.$$

(3.151)

From Equation 3.134 we recognize that the above expression is also the coefficient of the $(N–1)$th power term of the numerator polynomial $M(s)$ of A_{11}. Thus, by setting this coefficient to 1 in the numerator polynomial of Equation 3.147 and solving for μ_i, we obtain

$$\mu_i^2 = \frac{2(q_{N-1} - r_{N-1})}{1 + r_N}.$$

(3.152)

For filter designs based on symmetric CMR networks (i.e., $S_{11} = S_{22}$), we can set $\mu_o = \mu_i$.

Knowledge of μ_i allows us to determine the polynomials $M(s)$ and $D(s)$ of A_{11} as defined in Equation 3.147. Next, performing partial fraction expansions of the rational functions M/D and P/D, we can express the results as

$$\frac{M(s)}{D(s)} = \sum_{k=1}^{N} \frac{\xi_k^{(11)}}{s - \rho_k}, \tag{3.153}$$

$$\frac{P(s)}{D(s)} = \sum_{k=1}^{N} \frac{\xi_k^{(21)}}{s - \rho_k}, \tag{3.154}$$

where ρ_k are the poles and $\xi_k^{(11)}$ and $\xi_k^{(21)}$ are the residues of the respective rational function. Comparing the above expressions to Equations 3.134 and 3.135 shows that the eigenvalues of the matrix $\mathbf{M}$ are given by $\lambda_k = j\rho_k$. The elements $W_{1,k}$ and $W_{N,k}$ of the first and last rows of the matrix $\mathbf{W}$ are computed from the residues $\xi_k^{(11)}$ and $\xi_k^{(21)}$ as follows (Cameron 1999):

$$W_{1,k} = \left| \xi_k^{(11)} \right|^{1/2}, \tag{3.155}$$

$$W_{N,k} = \mathrm{sgn}\left\{ \mathrm{Im}\left[\xi_k^{(21)} \right] \right\} W_{1,k}, \tag{3.156}$$

where the sgn function returns the sign of its argument. The remaining rows of the matrix $\mathbf{W}$ can be obtained by Gram–Schmidt orthonormalization. Finally the energy coupling matrix $\mathbf{M}$ is obtained from $\mathbf{M} = \mathbf{W}\Lambda\mathbf{W}^{\mathrm{T}}$, where Λ is the diagonal matrix containing the eigenvalues λ_k.

The above synthesis procedure typically yields a full coupling matrix $\mathbf{M}$ which corresponds to a coupling topology that may not be realizable due to physical layout constraints. For example, the coupling topology may require a microring to be coupled to too many other microrings so that adjacent resonators would touch or overlap each other. It may also contain quadruplets with cross-couplings as shown in Figure 3.21a, or coupling loops with an odd number of microrings as shown in Figure 3.21b. The latter structures are undesirable since they lead to coupling between counter-propagating modes in the microrings and result in a reflected wave at the input port. In general, it is possible to convert an unrealizable—or undesirable—coupling configuration to a simpler and realizable one by applying similarity transformations such as Jacobi rotations to the coupling matrix $\mathbf{M}$ without disturbing its eigenvalues. Each Jacobi rotation yields a new coupling matrix $\mathbf{M'}$ according to

$$\mathbf{M'} = \mathbf{R}(\theta_r) \cdot \mathbf{M} \cdot \mathbf{R}^T(\theta_r), \tag{3.157}$$

where $\mathbf{R}(\theta_r)$ is an $N \times N$ rotation matrix and θ_r is the rotation angle chosen to annihilate an undesirable coupling element in the original coupling matrix $\mathbf{M}$ (Cameron 1999, 2003). We also require that the first and last rows of $\mathbf{R}$ to have 1s on the diagonal and zeros as the off-diagonal elements so as not to disturb the first and last microrings, since these are fixed by the input and output port locations. In general, we can always reduce a given coupling matrix to the so-called $2 \times m$ canonical form, which has the fewest number of coupling elements. The canonical coupling topology has the form of a folded microring array, such as the one shown in Figure 3.22a for a six-ring network. A procedure for reducing a given coupling matrix $\mathbf{M}$ to the canonical coupling form can be found in Cameron (1999).

In practical microring filter design, we typically perform the CMR synthesis procedure for a prototype filter with normalized center frequency $\omega_0 = 1$ rad/s and cutoff frequency $\omega_c = 1$ rad/s (bandwidth of 2 rad/s). The energy coupling coefficients are then scaled to obtain a filter with a specified bandwidth B (rad/s) according to the relations $\tilde{\mu}_{(i,o)} = \mu_{(i,o)}\sqrt{B/2}$ and $\tilde{\mu}_{i,j} = \mu_{i,j}(B/2)$, where the tilde sign denotes parameters of the new bandpass filter with bandwidth B. In the physical implementation of the microring filter, the energy coupling coefficients are converted to the field coupling

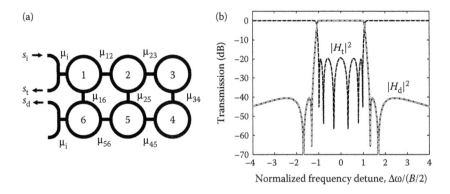

Figure 3.22   (a) Schematic of a 6th-order canonical CMR filter and (b) target filter responses (gray dashed lines) and spectral responses of the synthesized canonical CMR filter (black solid lines).

coefficients for microring resonators with a specified FSR (in Hertz) using the relations $\kappa_{(i,o)} = \tilde{\mu}_{(i,o)}/\sqrt{FSR}$ and $\kappa_{i,j} = \tilde{\mu}_{i,j}/FSR$.

We illustrate the above synthesis procedure with the design of a sixth-order pseudo-elliptic optical filter with 0.05 dB ripple in the passband and −40 dB stopband rejection. Choosing a design with four transmission zeros, we employ a suitable filter approximation method (e.g., Ellis 1994) to obtain a sixth-order transfer function that meets the above specifications. The polynomials of the drop port and through port transfer functions of the prototype filter (with a bandwidth of 2 rad/s) are given by

$$P(s) = 0.1128(s^4 + 4.5861s^2 + 4.9410),$$

$$R(s) = s^6 + 1.6716s^4 + 0.7452s^2 + 0.0589,$$

$$Q(s) = s^6 + 1.9528s^5 + 3.5783s^4 + 3.8036s^3 + 3.1742s^2$$
$$+ 1.7013s + 0.5606. \tag{3.158}$$

The filter has six poles located at $p_k = \{-0.0744 \pm j1.0609, -0.2961 \pm j0.9154, -0.6059 \pm j0.4103\}$. The drop port transfer function has four transmission zeros located on the imaginary axis at $z_k = \{\pm j1.6900, \pm j1.3153\}$, while the through port transfer function has six zeros on the imaginary axis at $z_k = \{\pm j0.3166, \pm j0.7834, \pm j0.9786\}$. The target spectral responses of the filter are shown by the gray dashed lines in Figure 3.22b.

In the microring filter design, we first use Equation 3.152 to compute the input and output coupling coefficients of the CMR filter to get $\mu_i = \mu_o = 1.3974$. Knowledge of μ_i and μ_o allows us to determine the polynomials $M(s)$ and $D(s)$ from Equations 3.147 and 3.148. Performing partial fraction expansions of the rational functions M/D and P/D, we obtain the poles and residues that are listed in Table 3.3. The first and last rows of the orthonormal matrix $\mathbf{W}$ are

Table 3.3   Poles and Residues of the Network Parameters A_{11} and A_{N1}

Poles ρ_k	$\xi_k^{(11)}$	$\xi_k^{(21)}$
$j1.1429$	0.0752	$-j0.1468$
$-j1.0507$	0.0752	$j0.1468$
$j1.0507$	0.1677	$j0.3275$
$-j1.0507$	0.1677	$-j0.3275$
$j0.4635$	0.2571	$-j0.5020$
$-j0.4635$	0.2571	$j0.5020$

next calculated using Equations 3.155 and 3.156 and the remaining rows are obtained by Gram–Schmidt orthonormalization. From $\mathbf{W}$ and the eigenvalue matrix Λ, we calculate the coupling matrix $\mathbf{M}$ to be (only the upper half is shown since the matrix is symmetric)

$$
\mathbf{M} = \begin{bmatrix}
0 & -0.2175 & 0.5516 & -0.5329 & -0.1957 & -0.0578 \\
  & -0.8264 & 0.0740 & 0.1033 & -0.5557 & -0.2897 \\
  &   & 0.5528 & -0.0439 & -0.1381 & 0.4749 \\
  &   &   & -0.4852 & -0.1671 & 0.4712 \\
  &   &   &   & 0.7589 & 0.3774 \\
  &   &   &   &   & 0
\end{bmatrix}.
$$

$$(3.159)$$

The above matrix corresponds to a coupling topology that is not physically realizable since it requires every microring in the network to be coupled to all the other microrings. By applying a series of matrix rotations as described in Cameron (1999), we can reduce the above coupling matrix to the canonical form

$$
\mathbf{M_C} = \begin{bmatrix}
0 & 0.8209 & 0 & 0 & 0 & 0.0578 \\
  & 0 & 0.5393 & 0 & 0.2840 & 0 \\
  &   & 0 & 0.7820 & 0 & 0 \\
  &   &   & 0 & -0.5393 & 0 \\
  &   &   &   & 0 & -0.8209 \\
  &   &   &   &   & 0
\end{bmatrix}.
$$

$$(3.160)$$

It is evident that the canonical coupling matrix has much fewer coupling elements than the original matrix in Equation 3.159. A schematic of the canonical CMR filter is shown in Figure 3.22a. Its spectral responses are shown by the black solid lines in Figure 3.22b, which are in good agreement with the target filter responses.

We observe in Equation 3.160 that all the diagonal elements of the canonical coupling matrix $\mathbf{M_C}$ are zero, implying that the microring resonators are synchronously tuned to the center frequency ω_0 of the filter passband. We also note that the coupling elements μ_{45} and μ_{56} are negative. In general, 2D CMR networks require negative coupling elements to realize transfer functions with transmission zeros on the $j\omega$-axis. One way to realize a negative coupling element is

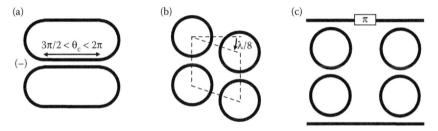

Figure 3.23 Schemes for realizing negative couplings in a CMR network: (a) micro-racetracks with coupling phase $\theta_c = k_c L$ such that $3\pi/2 < \theta_c < 2\pi$, (b) quadrupole coupling loop sheared by $\lambda/8$, and (c) cascaded microring doublets with inter-stage π-phase shift.

by evanescent coupling between two micro-racetracks with a long coupling length L such that $3\pi/2 < k_c L < 2\pi$, where k_c is the coupling strength, as shown in Figure 3.23a (Van 2007). Alternatively, negative coupling can be simulated in a loop-coupled microring quadruplet by shearing the loop by $(2n + 1)\lambda/8$, where n is an integer, as shown in Figure 3.23b (Popovic 2007, Tsay and Van 2012b). Finally, we note that the requirement for negative coupling elements is eliminated by using the microring ladder filter configuration in Section 3.6. Indeed, it can be shown that any CMR network with negative coupling coefficients can be decomposed into an array of cascaded microring doublets with all-positive coupling coefficients and differential π-phase shifts between the stages (Prabhu et al. 2008b). This is illustrated in Figure 3.22c for a quadrupole microring filter with a negative coupling element.

Although 2D CMR filters are generally more difficult to realize than serially coupled microring filters, attempts at demonstrating quadrupole 2D CMR filters with negative coupling using the sheared coupling loop configuration have been reported in SOI and SiN material systems (Popovic et al. 2008, Bachman et al. 2015). In particular, the measured transmission response of the silicon quadrupole pseudo-elliptic filter in Bachman et al. (2015) showed steep roll-offs and deep extinctions due to transmission nulls, thus providing clear evidence of negative coupling in the structure.

3.7.3 Power coupling analysis of 2D CMR networks

In the energy coupling analysis of 2D CMR networks, we implicitly implicitlyassume that the field amplitude in each microring resonator is uniformly distributed around the microring. This assumption, which holds for weak coupling, allows us to simplify the analysis by considering only the energy stored in each resonator (given by $|a_i(t)|^2$). Since the energy coupling formalism does not take into account the phase or amplitude variation of the signal in the microring waveguide, the response of the CMR network does not depend on the exact locations of the coupling junctions on the microrings.

For strongly coupled microring resonators, the field amplitudes before and after a coupling junction can differ significantly, and a more accurate analysis based on the coupling of power in space must be used to describe the device response. Such an analysis will also take into account the exact locations of the coupling junctions on the microrings. In this section, we develop a general method for analyzing 2D CMR networks using the power coupling formalism. We will show that any 2D microring coupling topology can be described in terms of a direct coupling matrix plus a commutator matrix responsible for nonadjacent resonator couplings. The latter term, which is neglected in the energy coupling approximation, can give rise to prominent resonance features that are not observed in weakly coupled CMR networks (Tsay and Van 2011b).

We consider again a 2D CMR network consisting of N microrings arranged in a square lattice as shown in Figure 3.24a. To simplify the analysis, we assume that the microrings are identical with the same radius R. Coupling between two adjacent microrings i and j is denoted by the field coupling coefficient $\kappa_{i,j}$. The input and output coupling coefficients are denoted by κ_i and κ_o, respectively. The signals at the input, through, drop, and add ports are labeled s_i, s_t, s_d, and s_a, respectively, which are normalized with respect to power. In each microring i, we label the amplitude of the circulating wave after each coupling junction as a_i, b_i, c_i, d_i, in the order of the direction of wave propagation, as shown in Figure 3.24a. These waves are also normalized with respect to power. The labeling order, $a \rightarrow b \rightarrow c \rightarrow d$, is also followed for signal paths crossing into adjacent microrings. For example, in Figure 3.24a, the field a_1 in ring 1 is coupled to field b_2 in ring 2 via coupling coefficient κ_{12}, which is then coupled to field c_{2m-1} in ring $2m-1$ via coupling coefficient $\kappa_{2,2m-1}$, and so on. This order of field labeling yields coupling matrices with simple structures in the analysis below.

To facilitate the analysis of the CMR network, we transform the structure into an equivalent array of coupled straight waveguides (Tsay and Van 2011b). This is achieved by "cutting" each microring

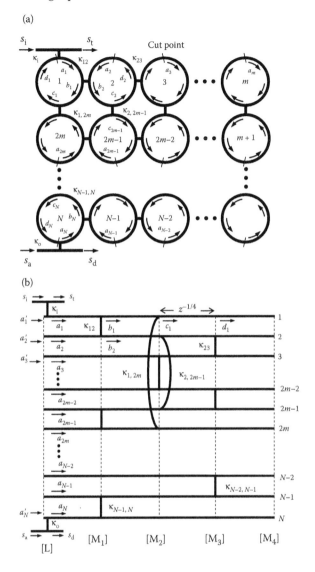

Figure 3.24 (a) Schematic of a 2D CMR network consisting of N microring resonators and (b) equivalent unfolded coupled waveguide array of the CMR structure in (a).

at the point just before the field a_i is defined and unfolding it into a straight waveguide while keeping track of the coupling junctions between adjacent microrings. In this manner, the CMR structure in Figure 3.24a is transformed into an equivalent "unfolded" coupled-waveguides configuration shown in Figure 3.24b, where a connection between two waveguides denotes a coupling junction.

The unfolded structure resembles an array of coupled FP waveguide cavities, except that the facets have periodic instead of reflective boundary conditions so that the waves travel only in the forward direction.

The transfer functions of the CMR network can now be obtained by analyzing the unfolded coupled waveguide array using the transfer matrix method. We can view the waveguide array as consisting of four cascaded sections, each representing a phase delay of a quarter ring, $e^{-j\phi_{rt}/4} = z^{-1/4}$, where ϕ_{rt} is the microring round-trip phase. We denote the fields in each section by the arrays $a, b, c,$ and d, where

$$a = \begin{bmatrix} a_1, & a_2, & \cdots & a_{N-1}, & a_N \end{bmatrix}^T, \tag{3.161}$$

with similar expressions for $b, c,$ and d. The sections are connected to each other by a set of coupling junctions that can be described by the four transfer matrices $\mathbf{M}_1, \mathbf{M}_2, \mathbf{M}_3,$ and $\mathbf{M}_4$. These $N \times N$ matrices are symmetric with the property that if there is coupling between waveguides i and j at junction k ($k = 1$–4) with coupling coefficient $\kappa_{i,j}$, then

$$\begin{aligned} \mathbf{M}_k(i,i) &= \mathbf{M}_k(j,j) = \tau_{i,j}, \\ \mathbf{M}_k(i,j) &= \mathbf{M}_k(j,i) = -j\kappa_{i,j}, \end{aligned} \tag{3.162}$$

where $\tau_{i,j}^2 + \kappa_{i,j}^2 = 1$. If waveguide i is uncoupled at junction k, then $\mathbf{M}_k(i,i) = 1$. Specifically, with the field labeling as shown in Figure 3.24a, the matrix $\mathbf{M}_1$ has the block-diagonal form

$$\mathbf{M}_1 = \begin{bmatrix} \mathbf{K}_{12} & & & & \\ & \mathbf{K}_{34} & & & \\ & & \ddots & & \\ & & & \mathbf{K}_{i,j} & \\ & & & & \ddots \end{bmatrix}, \tag{3.163}$$

where $\mathbf{K}_{i,j}$ is the 2×2 coupling matrix associated with the coupling junction between microrings i and j,

$$\mathbf{K}_{i,j} = \begin{bmatrix} \tau_{i,j} & -j\kappa_{i,j} \\ -j\kappa_{i,j} & \tau_{i,j} \end{bmatrix}. \tag{3.164}$$

The other coupling matrices $\mathbf{M}_k$ can also be transformed to the above block-diagonal form through some suitable permutations. Using these transfer matrices, we can relate the field arrays through the expressions $b = z^{-1/4}\mathbf{M}_1 a$, $c = z^{-1/4}\mathbf{M}_2 b$, $d = z^{-1/4}\mathbf{M}_3 c$, and $a' = z^{-1/4}\mathbf{M}_4 d$, where a' is the field array defined just before the input and output coupling junctions, as indicated in Figure 3.24b. These relations can be combined to give

$$a' = z^{-1}\mathbf{M}_4\mathbf{M}_3\mathbf{M}_2\mathbf{M}_1 a \equiv z^{-1}\mathbf{M}a, \tag{3.165}$$

where $\mathbf{M}$ represents the ring-to-ring coupling matrix of the CMR network. At the left boundary of the coupled waveguide array, the field arrays a and a' are also related by

$$a = \mathbf{L}a' + s, \tag{3.166}$$

where $s = [-j\kappa_i s_i, 0, \ldots 0, -j\kappa_o s_a]^T$ is the input field array and $\mathbf{L}$ is a diagonal matrix representing the bus-to-ring couplings,

$$\mathbf{L} = \mathrm{diag}\big[\tau_i, \ 1, \ \cdots \ 1, \ \tau_o\big], \tag{3.167}$$

with $\tau_{(i,o)} = \sqrt{1 - \kappa_{(i,o)}^2}$. Upon substituting Equation 3.165 into Equation 3.166, we obtain

$$(\mathbf{I} - z^{-1}\mathbf{L}\mathbf{M})a = s, \tag{3.168}$$

where $\mathbf{I}$ is the $N \times N$ identity matrix. The above equation shows that the spectral responses of the field amplitudes in a CMR network are characterized by a field coupling matrix $\mathbf{M}$, which is determined by the coupling topology, and a ring-to-bus coupling matrix $\mathbf{L}$ representing extrinsic losses due to coupling to the bus waveguides. Internal loss in the microrings can be accounted for by letting $z^{-1} = a_{rt}e^{-j\phi_{rt}}$, where a_{rt} is the round-trip amplitude attenuation. We also note that Equation 3.168 has a somewhat similar form to Equation 3.118 obtained using the energy coupling formalism. Indeed, it can be shown that Equation 3.118 can be derived from the more general Equation 3.168 under the weak coupling condition (Tsay and Van 2011b).

The field coupling equation in Equation 3.168 can be solved by diagonalizing the matrix product $\mathbf{L}\mathbf{M}$ in the form $\mathbf{L}\mathbf{M} = \mathbf{Q}\mathbf{D}\mathbf{Q}^{-1}$, where $\mathbf{D}$ is a diagonal matrix containing the eigenvalues of $\mathbf{L}\mathbf{M}$ and $\mathbf{Q}$ is the corresponding eigenvector matrix. Making this substitution in Equation 3.168 and solving for a, we get

$$a = \mathbf{Q}(\mathbf{I} - z^{-1}\mathbf{D})^{-1}\mathbf{Q}^{-1}s. \tag{3.169}$$

Assuming that only the input port is excited ($s_i \neq 0$, $s_a = 0$), we obtain from Equation 3.169 the field amplitude a_i in microring i as follows:

$$a_i = -j\kappa_i s_i \sum_{k=1}^{N} \frac{Q_{i,k} Q_{k,1}^{-1}}{1 - z^{-1}\lambda_k}, \tag{3.170}$$

where λ_k are the diagonal entries of $\mathbf{D}$, and $Q_{i,k}$ and $Q_{k,1}^{-1}$ are elements of the matrices $\mathbf{Q}$ and $\mathbf{Q}^{-1}$, respectively.

The transfer functions at the through port and drop port of the CMR network can be obtained by relating the output signals s_t and s_d to the fields a_1 and a_N in microrings 1 and N. At the coupling junction between the input waveguide and microring 1, we have the relations

$$s_t = \tau_i s_i - j\kappa_i a_1', \tag{3.171}$$

$$a_1 = -j\kappa_i s_i + \tau_i a_1'. \tag{3.172}$$

Eliminating a_1' from the above equations gives

$$s_t = \frac{1}{\tau_i}(s_i - j\kappa_i a_1). \tag{3.173}$$

Using Equation 3.170 for a_1, we obtain the through port transfer function of the CMR filter as follows:

$$H_t(z^{-1}) = \frac{s_t}{s_i} = \frac{1}{\tau_i}\left(1 - \kappa_i^2 \sum_{k=1}^{N} \frac{Q_{1,k} Q_{k,1}^{-1}}{1 - z^{-1}\lambda_k}\right). \tag{3.174}$$

To determine the drop port response, we set $s_a = 0$ so that at the output coupling junction of microring N, we have $s_d = -j\kappa_o a_N'$ and $a_N = \tau_o a_N'$. These two relations are combined to give $s_d = -j\kappa_o a_N/\tau_o$. Using Equation 3.170 for a_N, we obtain, for the drop port transfer function of the CMR filter,

$$H_d(z^{-1}) = \frac{s_d}{s_i} = -\frac{\kappa_i \kappa_o}{\tau_o} \sum_{k=1}^{N} \frac{Q_{N,k} Q_{k,1}^{-1}}{1 - z^{-1}\lambda_k}. \tag{3.175}$$

Equations 3.174 and 3.175 show that, in the absence of resonator loss, the poles of the CMR filter are given by the inverse of the eigenvalues $(1/\lambda_k)$ of the matrix product $\mathbf{LM}$. We also note that the zeroth-order term of the numerator polynomial of H_d is zero,

$$-\frac{\kappa_i \kappa_o}{\tau_o} \sum_{k=1}^{N} Q_{N,k} Q_{k,1}^{-1} = 0, \tag{3.176}$$

since the summation is the product of row N of $\mathbf{Q}$ and column 1 of $\mathbf{Q}^{-1}$. Thus we can write the drop port transfer function in the form

$$H_d(z^{-1}) = \frac{z^{-1} P(z^{-1})}{Q(z^{-1})}, \tag{3.177}$$

where P is a polynomial of maximum degree $N-2$. Equation 3.177 indicates that excluding the trivial zero at $z^{-1} = 0$, the drop port transfer function of the CMR network has a maximum of $N-2$ transmission zeros. We recall that this was also shown to be the case for the device transfer function in the s-domain (Equation 3.124).

If we take the absolute square of each term in the summation of Equation 3.175, we obtain an expression of the form

$$\left| \frac{Q_{N,k} Q_{k,1}^{-1}}{1 - \lambda_k e^{-j\phi_{rt}}} \right|^2 \propto \frac{1}{1 + F_k \sin^2(\phi)}, \tag{3.178}$$

where $\phi = (\phi_{rt} - \theta_k)/2$, θ_k is the angle of λ_k, and $F_k = 4|\lambda_k|/(1 - |\lambda_k|^2)$. The above expression has the form of an Airy function. Thus the response of a CMR network can be regarded as consisting of a sum of N Airy resonances.[*] In contrast, energy coupling analysis shows that the response of a weakly coupled CMR network consists of a sum of N Lorentzian resonances.

We also saw in Section 3.7.1 that in the energy coupling formalism, the coupling matrix $\mathbf{M}$ of a 2D CMR network specifies the direct couplings between adjacent microring resonators, that is, the element $\mathbf{M}(i, j)$ specifies the energy coupling coefficient $\mu_{i,j}$ between neighbor microrings i and j. In contrast, the structure of the complex coupling matrix $\mathbf{M}$ in the power coupling formalism is more complicated. To deduce the physical meaning of the field coupling matrix $\mathbf{M}$, we first look at the individual coupling matrices

---

[*] See also Footnote on page 138.

$\mathbf{M}_k$ of the coupled waveguide array. Each of these matrices, through some suitable permutation, can be expressed in the block-diagonal form of Equation 3.163, in which the block matrices are the junction coupling matrices $\mathbf{K}_{i,j}$. Since $\mathbf{K}_{i,j}$ are circular matrices,* the coupling matrices $\mathbf{M}_k$ are also circular, which means that they can be expressed in the form

$$\mathbf{M}_k = \mathbf{W}_k \mathbf{\Lambda}_k \mathbf{W}_k^T = e^{j\Psi_k}, \tag{3.179}$$

where $\mathbf{\Lambda}_k$ is a diagonal matrix containing the eigenvalues of $\mathbf{M}_k$ and $\mathbf{W}_k$ is a real matrix containing the corresponding eigenvectors. The eigenvalues of $\mathbf{M}_k$ are given by

$$\mathbf{\Lambda}_k(i,i) = \mathbf{\Lambda}_k^*(j,j) = \tau_{i,j} - j\kappa_{i,j} = e^{j\theta_{i,j}}, \tag{3.180}$$

where $\theta_{i,j} = -\tan^{-1}(\kappa_{i,j}/\tau_{i,j})$ is defined as the coupling angle between waveguides i and j. If waveguide i of section k is not coupled to any other waveguide, then $\mathbf{\Lambda}_k(i,i) = 1$. The matrix Ψ_k in Equation 3.179 is real and symmetric with zeros on the diagonal. The off-diagonal elements are also zero except that if there is direct coupling between waveguides i and j, then $\Psi_k(i,j) = \Psi_k(j,i) = \theta_{i,j}$. We call Ψ_k the coupling-angle matrix of section k of the coupled waveguide array.

Using Equation 3.179, we can express the total coupling matrix $\mathbf{M}$ of the CMR network as

$$\mathbf{M} = \mathbf{M}_4 \mathbf{M}_3 \mathbf{M}_2 \mathbf{M}_1 = e^{j\Psi_4} e^{j\Psi_3} e^{j\Psi_2} e^{j\Psi_1}. \tag{3.181}$$

Since the matrices $\mathbf{M}_k$ do not commute in general, we use the Baker–Campbell–Hausdorff formula to write the product of the matrix exponentials in Equation 3.181 as

$$\mathbf{M} = \exp\left[j(\Psi_1 + \Psi_2 + \Psi_3 + \Psi_4) + j\mathbf{X}\right] = \exp(j\Psi + j\mathbf{X}). \tag{3.182}$$

In the above equation, the matrix sum $\Psi = \Psi_1 + \Psi_2 + \Psi_3 + \Psi_4$ is a symmetric matrix with zeros on the diagonal. Its off-diagonal elements are given by the coupling angle between microrings

---

* A circular matrix $\mathbf{M}$ is a matrix that can be expressed in the form $\mathbf{M} = e^{j\psi}$, where ψ is a real matrix.

i and j, $\Psi(i,j) = \Psi(i,j) = \theta_{i,j}$. The matrix $\mathbf{X}$ denotes the sum of all nested commutators,

$$\mathbf{X} = \sum_m^\infty P_m(\Psi_1, \Psi_2, \Psi_3, \Psi_4), \tag{3.183}$$

where P_m represents polynomials of the commutators of the matrices Ψ_k. Equation 3.182 shows that, in general, a 2D CMR network is characterized by a coupling-angle matrix Ψ plus a commutation matrix $\mathbf{X}$. The matrix Ψ accounts for the direct couplings between adjacent microring resonators and has the same form as the energy coupling matrix $\mathbf{M}$ in the energy coupling formulation, except that its (i, j) element is given by the coupling angle $\theta_{i,j}$ instead of the energy coupling coefficient $\mu_{i,j}$. As in the case of the energy coupling matrix, the direct coupling-angle matrix Ψ explicitly reflects the 2D coupling topology of the CMR network (i.e., the coupling topology can be deduced from Ψ).

The commutation matrix $\mathbf{X}$ in Equation 3.182 accounts for the effective couplings arising from all the indirect coupling paths between two microrings that may or may not be neighbors. It is in general a full but symmetric matrix, with alternating diagonal bands of real and pure imaginary elements. In the limit of weak coupling, the indirect coupling matrix can be neglected ($\mathbf{X} \approx 0$)[*] so that $\mathbf{M}$ can be approximated by just the direct coupling term, $\mathbf{M} \approx e^{j\Psi}$. Indeed, it can be shown that this is the approximation made in the energy coupling formulation of 2D CMRs (Tsay and Van 2011b). We can thus conclude that the energy coupling formulation does not take into account the effect of the indirect couplings. For strongly coupled CMR networks, the indirect coupling term can give rise to resonance features that are not observed in weakly coupled systems (Tsay and Van 2011b).

Due to the noncommutative nature of the coupling matrices $\mathbf{M}_k$, a method for exactly synthesizing CMR filters of general 2D coupling topology in the z-domain has not been found. An approximate method that assumes that the matrices $\mathbf{M}_k$ are commutative is

---

[*] The commutation matrix $\mathbf{X}$ also vanishes for microring coupling topologies whose matrices $\mathbf{M}_k$ do commute. One such special case is a quadruplet with identical coupling coefficients $\kappa_{i,j} = \kappa$. For these structures, all the indirect coupling paths cancel themselves out and the coupling matrix is given by only the direct coupling terms.

given in Tsay and Van (2012a). On the other hand, exact synthesis can be achieved if one assumes *a priori* the coupling topology of the CMR network. This is the approach adopted in Tsay and Van (2011a), where an exact method for synthesizing CMR filters in the canonical $2 \times m$ coupling topology is developed based on the network order reduction approach.

In addition to their applications in realizing advanced optical filters, 2D CMR networks can also be used to construct photonic analogs of many 2D electronic systems, allowing many interesting properties of these systems to be studied and explored in the optical domain. For example, the clockwise and counterclockwise modes in a CMR square lattice exhibit properties similar to the cyclotron orbits of electrons in an atomic lattice subject to a magnetic field. In particular, it has been shown that, with proper choice of the coupling phases between adjacent microrings, the CMR lattice behaves like a 2D atomic lattice subject to a uniform perpendicular magnetic field to the lattice (Hafezi et al. 2011, 2013, Liang and Chong 2013). In a finite CMR lattice, topologically protected edge states can emerge which are immune to back scattering caused by imperfections in the lattice. It has been suggested that such a CMR lattice can be used to realize photonic integrated devices such as optical delay lines that are robust to fabrication variations (Hafezi et al. 2011).

3.8   Summary

In this chapter we developed general techniques for analyzing and designing coupled microring optical filters. Several common coupling configurations in 1D and 2D were considered and the type of filter transfer functions realizable by each configuration was derived. Given a set of filter specifications, the most suitable choice of the microring filter to use is typically determined by the number of resonators required to synthesize the desired spectral response, the complexity of the resulting device configuration, and the sensitivity of the design to parameter variations. Currently, one of the biggest challenges to the practical realization of high-order microring filters is the issue of resonance mismatches caused by fabrication imperfections. As advances in fabrication techniques allow for better control of device dimensions and more robust methods are developed for post-fabrication fine-tuning of the microring resonances, it is expected that increasingly more complex microring device architectures can be realized for advanced integrated optics filter applications.

References

Antoniou, A. 1993. *Digital Filters: Analysis, Design, and Applications.* New York: McGraw-Hill.

Atia, A. E., Williams, A. E., Newcomb, R. W. 1974. Narrow-band multiple-coupled cavity synthesis. *IEEE Trans. Circuits Syst.* CAS-21(5): 649–655.

Bachman, D., Tsay, A., Van, V. 2015. Negative coupling and coupling phase dispersion in a silicon quadrupole micro-racetrack resonator. *Opt. Express* 23(15): 20089–20095.

Barwicz, T., Popovic, M. A., Rakich, P. T., Watts, M. R., Haus, H. A., Ippen, E. P., Smith, H. I. 2004. Microring-resonator-based add-drop filters in SiN: Fabrication and analysis. *Opt. Express* 12(7): 1437–1442.

Cameron, R. J. 1999. General coupling matrix synthesis methods for Chebyshev filtering functions. *IEEE Trans. Microwave Theory Tech.* 47: 433–442.

Cameron, R. J. 2003. Advanced coupling matrix synthesis techniques for microwave filters. *IEEE Trans. Microwave Theory Tech.* 51: 1–10.

Chremmos, I., Uzunoglu, N. 2008. Modes of the infinite square lattice of coupled microring resonators. *J. Opt. Soc. Am. A* 25(12): 3043–3050.

Cooper, M. L., Gupta, G., Green, W. M., Assefa, S., Xia, F., Vlasov, Y. A., Mookherjea, S. 2010. 235-Ring coupled-resonator optical waveguides. In Conference on Lasers and Electro-Optics, Optical Society of America, San Jose, CA, paper CTuHH3.

Grover, R., Van, V., Ibrahim, T. A., Absil, P. P., Calhoun, L. C., Johnson, F. G., Hryniewicz, J. V., Ho, P. T. 2002. Parallel-cascaded semiconductor microring resonators for high-order and wide-FSR filters. *J. Lightwave Technol.* 20(5): 900–905.

Ellis, M. G. 1994. *Electronic Filter Analysis and Synthesis.* Norwood, MA: Artech House.

Hafezi, M., Demler, E. A., Lukin, M. D., Taylor, J. M. 2011. Robust optical delay lines with topological protection. *Nat. Phys.* 7: 907–912.

Hafezi, M., Mittal, S., Fan, J., Migdall, A., Taylor, J. M. 2013. Imaging topological edge states in silicon photonics. *Nat. Photonics* 7: 1001–1005.

Haus, H. A. 1984. *Waves and Fields in Optoelectronics.* Englewood Cliffs, NJ: Prentice-Hall.

Heebner, J. E., Chak, P., Pereira, S., Sipe, J. E., Boyd, R. W. 2004. Distributed and localized feedback in microresonator sequences for linear and nonlinear optics. *J. Opt. Soc. Am. B* 21(10): 1818–1832.

Hill, K. O., Meltz, G. 1997. Fiber Bragg grating technology fundamentals and overview. *J. Lightwave Technol.* 15(8): 1263–1276.

Hryniewicz, J. V., Absil, P. P., Little, B. E., Wilson, R. A., Ho, P.-T. 2000. Higher order filter response in coupled microring resonators. *IEEE Photonics Technol. Lett.* 12(3): 320–322.

Jinguji, K., Oguma, M. 2000. Optical half-band filters. *J. Lightwave Technol.* 18(2): 252–259.

Lam, H. Y.-F. 1979. *Analog and Digital Filters: Design and Realization.* Englewood Cliffs, NJ: Prentice-Hall.

Liang, G. Q., Chong, Y. D. 2013. Optical resonator analog of a two-dimensional topological insulator. *Phys. Rev. Lett.* 110: 203904.

Liew, H. L., Van, V. 2008. Exact realization of optical transfer functions with symmetric transmission zeros using the double-microring ladder architecture. *J. Lightwave Technol.* 26: 2323–2331.

Little, B. E., Chu, S. T., Haus, H. A., Foresi, J., Laine, J.-P. 1997. Microring resonator channel dropping filters. *J. Lightwave Technol.* 15(6): 998–1005.

Little, B. E., Chu, S. T., Hryniewicz, J. V., Absil, P. P. 2000. Filter synthesis for periodically coupled microring resonators. *Opt. Lett.* 25(5): 344–346.

Little, B. E., Chu, S. T., Absil, P. P., Hryniewicz, J. V., Johnson, F. G., Seiferth, F., Gill, D., Van, V., King, O., Trakalo, M. 2004. Very high order microring resonator filters for WDM applications. *IEEE Photonics Technol. Lett.* 16(10): 2263–2265.

Madsen, C. K. 1998. Efficient architectures for exactly realizing optical filters with optimum bandpass design. *IEEE Photonics Technol. Lett.* 10: 1136–1138.

Madsen, C. K., Lenz, G. 1998. Optical all-pass filters for phase response design with applications for dispersion compensation. *IEEE Photonics Technol. Lett.* 10(7): 994–996.

Madsen, C. K., Lenz, G., Bruce, A. J., Cappuzzo, M. A., Gomez, L. T., Scotti, R. E. 1999. Integrated all-pass filters for tunable dispersion and dispersion slope compensation. *IEEE Photonics Technol. Lett.* 11(12): 1623–1625.

Madsen, C. K., Zhao, J. H. 1999. *Optical Filter Design and Analysis: A Signal Processing Approach.* New York: John Wiley & Sons Inc.

Masilamani, A. P., Van, V. 2012. Design and realization of a two-stage microring ladder filter in silicon-on-insulator. *Opt. Express* 20(22): 24708–24713.

Martinez, H.G., Parks, T.W. 1978. Design of recursive digital filters with optimum magnitude and attenuation poles on the unit circle. *IEEE Trans. Acoust. Speech Signal Process.* 26(2): 150–156.

Melloni, A., Morichetti, F., Martinelli, M. 2003. Linear and nonlinear pulse propagation in coupled resonator slow-wave optical structures. *Opt. Quantum Electron.* 35: 365–379.

Orta, R., Savi, P., Tascone, R., Trinchero, D. 1995. Synthesis of multiple-ring-resonator filters for optical systems. *IEEE Photonics Technol. Lett.* 7(12): 1447–1449.

Poon, J. K. S., Scheuer, J., Mookherjea, S., Paloczi, G. T., Huang, Y., Yariv, A. 2003. Matrix analysis of microring coupled-resonator optical waveguides. *Opt. Express* 12(1): 90–103.

Poon, J. K. S., Scheuer, J., Xu, Y., Yariv, A. 2004. Designing coupled-resonator optical waveguide delay lines. *J. Opt. Soc. Am. B* 21(9): 1665–1673.

Popovic, M. A. 2007. Sharply-defined optical filters and dispersion-less delay lines based on loop-coupled resonators and 'negative' coupling. In *Conference on Lasers and Electro-Optics*, Optical Society of America, Baltimore, MD, paper CThP6.

Popovic, M. A., Barwicz, T., Rakich, P. T., Dahlem, M. S., Holzwarth, C. W., Gan, F., Socci, L. et al. 2008. Experimental demonstration of loop-coupled microring resonators for optimally sharp optical filters. In *Conference on Lasers and Electro-Optics*, Optical Society of America, San Jose, CA, paper CTuNND.

Prabhu, A. M., Liew, H. L., Van, V. 2008a. Experimental determination of coupled-microring filter parameters via pole-zero extraction. *Opt. Express* 16(9): 14588–14596.

Prabhu, A. M., Liew, H. L., Van, V. 2008b. Generalized parallel-cascaded microring networks for spectral engineering applications. *J. Opt. Soc. Am. B* 25: 1505–1514.

Prabhu, A. M., Van, V. 2008. Predistortion techniques for synthesizing coupled microring filters with loss. *Opt. Commun.* 281: 2760–2767.

Tsay, A., Van, V. 2011a. A method for exact synthesis of 2xN coupled microring resonator networks. *IEEE Photonics Technol. Lett.* 23(23): 1778–1780.

Tsay, A., Van, V. 2011b. Analytic theory of strongly-coupled microring resonators. *IEEE J. Quantum Electron.* 47(7): 997–1005.

Tsay, A., Van, V. 2012a. Field coupling method for the direct synthesis of 2-D microring resonator networks. *IEEE J. Quantum Electron.* 48(10): 1314–1321.

Tsay, A., Van, V. 2012b. Analysis of coupled microring resonators in sheared lattices. *IEEE Photonics Technol. Lett.* 24(18): 1625–1627.

Van, V., Little, B. E., Chu, S. T., Hryniewicz, J. V. 2004. Micro-ring resonator filters. In *The 17th Annual Meeting of the IEEE Lasers and Electro-Optics Society*, Puerto Rico, Vol. 2, pp. 571–572.

Van, V. 2006. Circuit-based method for synthesizing serially coupled microring filters. *J. Lightwave Technol.* 24(7): 2912–2919.

Van, V. 2007. Synthesis of elliptic optical filters using mutually coupled microring resonators. *J. Lightwave Technol.* 25(2): 584–590.

Van, V. 2009. Canonic design of parallel cascades of symmetric two-port microring networks. *J. Lightwave Technol.* 27(2): 4870–4877.

Xia, F., Rooks, M., Sekaric, L., Vlasov, Y. 2007a. Ultra-compact high order ring resonator filters using submicron silicon photonic wires for on-chip optical interconnects. *Opt. Express* 15: 11934–11941.

Xia, F., Sekaric, L., Vlasov, Y. 2007b. Ultracompact optical buffers on a silicon chip. *Nat. Photonics* 1(1): 65–71.

Yariv, A., Xu, Y., Lee, R. K., Scherer, A. 1999. Coupled resonator optical waveguide: A proposal and analysis. *Opt. Lett.* 24: 711–713.

Yeh, P., Yariv, A., Hong, C.-S. 1977. Electromagnetic propagation in periodic stratified media. I. General theory. *J. Opt. Soc. Am.* 67(4): 423–438.

CHAPTER 4

Nonlinear Optics Applications of Microring Resonators

This chapter provides an overview of important nonlinear optical processes in microring resonators and their applications. In a high-Q microring resonator, the large buildup of light intensity at a resonance can lead to strongly enhanced nonlinear optical effects. These effects have been exploited for a wide range of applications, including all-optical switching, photonic logic operations, wavelength conversion, and parametric amplification. In this chapter we will focus mainly on intensity-dependent nonlinear effects, such as those arising from Kerr nonlinearity in dielectric and semiconductor materials, although the models and techniques developed in the chapter can also be applied to microring devices possessing second-order nonlinearity. In addition, in semiconductors such as silicon, significant two-photon absorption (TPA) may also occur, which results in a large concentration of free carriers being generated in the device. The effects of free carrier dispersion (FCD) and free carrier absorption (FCA) on the response of the microring device can be significant. We will also examine these free carrier-induced nonlinear effects in this chapter.

We will begin in Section 4.1 with a brief review of optical nonlinearity in materials and the formalisms for treating nonlinear wave propagation in optical waveguides. Section 4.2 will examine the effects of self-phase modulation (SPM) in a microring resonator, which at high powers can lead to the phenomena of bistability and self-pulsation. The application of microring resonators for all-optical switching will be discussed in Section 4.3. In Section 4.4, we will study the parametric processes of wavelength conversion and optical amplification based on four-wave mixing (FWM) in a microring resonator.

4.1 Nonlinearity in Optical Waveguides

4.1.1 Intensity-dependent nonlinearity

At high optical intensities, the induced polarization in a material exhibits a nonlinear dependence on the electric field. In the frequency domain, this nonlinear dependence can be expressed in terms of the power series

$$P(\mathbf{r},\omega) = \varepsilon_0 \chi^{(1)}(\omega)E(\mathbf{r},\omega) + \varepsilon_0 \chi^{(2)}(\omega)E^2(\mathbf{r},\omega) + \varepsilon_0 \chi^{(3)}(\omega)E^3(\mathbf{r},\omega) + \cdots,$$

(4.1)

where $E(\mathbf{r}, \omega)$ and $P(\mathbf{r}, \omega)$ are the Fourier transforms of the electric field $\mathcal{E}(\mathbf{r}, t)$ and polarization $\mathcal{P}(\mathbf{r}, t)$, and $\chi^{(n)}$ is the nth-order susceptibility. The above expression assumes that the electric field is linearly polarized and the medium is isotropic. In the more general case, the susceptibilities must be expressed as tensors. Specifically, the nth-order polarization vector $\mathbf{P}^{(n)}(\mathbf{r}, \omega)$ is related to the applied electric field $\mathbf{E}(\mathbf{r}, \omega)$ via a susceptibility tensor of rank $n + 1$. For example, the third-order polarization is given by

$$\mathbf{P}^{(3)}(\mathbf{r},\omega) = \varepsilon_0 \boldsymbol{\chi}^{(3)}(\omega) \vdots \mathbf{E}(\mathbf{r},\omega)\mathbf{E}(\mathbf{r},\omega)\mathbf{E}(\mathbf{r},\omega),$$

(4.2)

where $\boldsymbol{\chi}^{(3)}$ is a fourth-rank tensor and $\vdots$ denotes the tensor product. For materials with inversion symmetry such as amorphous solids (e.g., SiO_2, SiN) and semiconductors with centrosymmetric crystalline structures (e.g., Si, Ge), all the even-order terms in Equation 4.1 vanish, leaving the third-order polarization as the dominant source of nonlinearity. On the other hand, in non-centrosymmetric semiconductors such as GaAs and InP, both the second-order and third-order nonlinearities are present. However, depending on the application, one nonlinear process typically dominates over the other in the material, so we can neglect the latter process. In this chapter we will deal mainly with nonlinear processes that involve only the third-order susceptibility.

A general definition of the third-order susceptibility assumes the medium is illuminated by three waves at frequencies ω_m, ω_n, and ω_o. According to Equation 4.2, these waves induce a nonlinear polarization at the frequency $\omega_{onm} = \omega_o + \omega_n + \omega_m$ whose ith component is given by (Boyd 2008)

$$P_i^{(3)}(\omega_{onm}) = \varepsilon_0 D \sum_{j=1}^{3}\sum_{k=1}^{3}\sum_{l=1}^{3} \chi_{ijkl}^{(3)}(\omega_{onm};\omega_o,\omega_n,\omega_m)E_l(\omega_o)E_k(\omega_n)E_j(\omega_m),$$

(4.3)

where $\{i, j, k, l\}$ represent the x, y, and z components of the fields. The factor D is called the degenerate factor and is equal to the number of distinct permutations of the frequencies ω_m, ω_n, and ω_o which add up to ω_{onm}. Two important processes arise from the nonlinear interactions of these waves: self-phase modulation (SPM) and cross-phase modulation (XPM). SPM refers to the nonlinear polarization induced by a wave at frequency ω_1 on itself. This process occurs when the frequencies $\{\omega_m, \omega_n, \omega_o\}$ are equal to $\{\omega_1, \omega_1, -\omega_1\}$, $\{\omega_1, -\omega_1, \omega_1\}$, or $\{-\omega_1, \omega_1, \omega_1\}$, all of which give $\omega_{onm} = \omega_1$. The degenerate factor D is thus equal to 3 for SPM. In XPM, the nonlinear polarization experienced by a wave at frequency ω_1 is caused by another wave at ω_2. There are six combinations of the frequencies ω_1 and ω_2 which give rise to XPM at $\omega_{onm} = \omega_1$, an example being $\{\omega_m, \omega_n, \omega_o\} = \{\omega_1, \omega_2, -\omega_2\}$, so the degenerate factor D is equal to 6 in this case.

For isotropic materials, the susceptibility tensor $\chi^{(3)}$ is considerably simplified because of symmetry considerations. In particular, if we assume that all the waves have the same linear polarization, then the SPM and XPM polarizations are also parallel to the fields and are given simply by

$$P_{NL}^{SPM}(\omega_1) = 3\varepsilon_0 \chi^{(3)} \, |E(\omega_1)|^2 \, E(\omega_1), \tag{4.4}$$

$$P_{NL}^{XPM}(\omega_1) = 6\varepsilon_0 \chi^{(3)} \, |E(\omega_2)|^2 \, E(\omega_1), \tag{4.5}$$

where $\chi^{(3)}$ is the $\chi_{xxxx}^{(3)}$ component of the susceptibility tensor. Note that the above expressions indicate that the effect of XPM is twice as large as that of SPM. The total polarization in the medium is the sum of the linear and nonlinear polarizations,

$$P = P_L + P_{NL} = \varepsilon_0 [\, \chi^{(1)} + D\chi^{(3)} \, |E|^2] \, E, \tag{4.6}$$

where $D = 3$ for SPM and $D = 6$ for XPM. From the total polarization we can define the total relative permittivity of the medium in the presence of third-order nonlinearity as

$$\varepsilon_r = \varepsilon_{r,L} + \varepsilon_{r,NL} = 1 + \chi^{(1)} + D\chi^{(3)} \, |E|^2 \,. \tag{4.7}$$

In the above expression, we identify $\varepsilon_{r,L} = 1 + \chi^{(1)}$ as the linear relative permittivity and $\varepsilon_{r,NL} = D\chi^{(3)}|E|^2$ as the nonlinear relative permittivity.

The third-order nonlinearity is often expressed in terms of an intensity-dependent refractive index defined as

$$n = n_0 + n_2 I, \tag{4.8}$$

where n_0 and n_2 are the linear and nonlinear refractive indices, respectively, and I is the time-averaged optical intensity. For a plane wave with the electric field expressed as

$$\mathcal{E}(t) = E(\omega)e^{j\omega t} + \text{c.c.},\tag{4.9}$$

the optical intensity is

$$I = n_0\varepsilon_0 c\langle\mathcal{E}^2(t)\rangle = 2n_0\varepsilon_0 c\,|E(\omega)|^2\,.\tag{4.10}$$

By equating n^2 to the total relative permittivity in Equation 4.7, we get

$$n^2 = (n_0 + n_2 I)^2 = 1 + \chi^{(1)} + D\chi^{(3)}\,|E|^2\,.\tag{4.11}$$

Neglecting the second-order term in n_2, we obtain $n_0 = \sqrt{1+\chi^{(1)}}$ for the linear refractive index and

$$n_2 = \frac{D\chi^{(3)}}{4n_0^2\varepsilon_0 c}\tag{4.12}$$

for the nonlinear index.

An alternative definition of the intensity-dependent refractive index is by means of the expression

$$n = n_0 + \bar{n}_2\langle\mathcal{E}^2(t)\rangle.\tag{4.13}$$

By equating $\bar{n}_2\langle\mathcal{E}^2(t)\rangle = n_2 I$ and making use of Equation 4.10, we obtain the following relationship between n_2 and $\bar{n}_2$:

$$\bar{n}_2 = n_0\varepsilon_0 c n_2 = \frac{D\chi^{(3)}}{4n_0}\,.\tag{4.14}$$

Table 4.1 lists the values of the third-order susceptibility $\chi^{(3)}$ and the SPM nonlinear index n_2 near the 1.55 μm wavelength for some common materials of interest in integrated optics.

4.1.2   Free carrier-induced nonlinear effects

In a semiconductor material whose energy bandgap is less than twice the photon energy ($E_g < 2\hbar\omega$), TPA can occur which results in the generation of free electrons and holes. These electrons and holes

Table 4.1   Third-Order Susceptibility and Nonlinear Refractive
Index Near $\lambda = 1.55$ µm of Some Common Integrated
Optics Waveguide Materials

Material	n_0	$\chi^{(3)}$ (m²/V²)	n_2 (cm²/W)	References
SiO₂ (fused silica)	1.45	1.9×10^{-22}	2.6×10^{-16}	Eggleton et al. (2008)
SiN	2.0	3.4×10^{-21}	2.4×10^{-15}	Ikeda et al. (2008)
Si	3.45	1.9×10^{-19}	4.5×10^{-14}	Dinu et al. (2003)
AlGaAs	3.14	1.3×10^{-19}	3.6×10^{-14}	Islam et al. (1992)
GaAs	3.47	6.8×10^{-19}	1.6×10^{-13}	Islam et al. (1992)

modify the medium's optical properties through the FCA and FCD
effects, leading to a nonlinear change in the absorption and refractive index of the medium. For example, at the telecommunication
wavelengths, considerable TPA can occur in silicon at moderate to
high power levels, which gives rise to nonnegligible free carrier-
induced nonlinear effects.

The effects of free carriers on the optical polarization can be
accounted for by defining a nonlinear susceptibility due to free
carriers, χ_{fc}, as

$$\chi_{fc} = 2n_0 \left(\Delta n_{fc} - j \frac{\Delta \alpha_{fc}}{2k} \right),$$ (4.15)

where $k = \omega/c$, n_0 is the linear refractive index, and Δn_{fc} and $\Delta \alpha_{fc}$
are the changes in the refractive index and absorption, respectively,
induced by the free carriers. These changes are related to the generated electron and hole densities, N_e and N_h, in the medium according to the relations

$$\Delta n_{fc} = \sigma_r^{(e)} N_e + \sigma_r^{(h)} N_h,$$ (4.16)

$$\Delta \alpha_{fc} = \sigma_a^{(e)} N_e + \sigma_a^{(h)} N_h,$$ (4.17)

where $\sigma_r^{(e,h)}$ and $\sigma_a^{(e,h)}$ are the refraction volume and absorption
cross section, and the superscripts (e) and (h) indicate the contributions from electrons and holes, respectively. Since TPA generates an
equal number of electrons and holes, we can set $N_e = N_h = N_{fc}$ in the
above relations to get

$$\Delta n_{fc} = \left[\sigma_r^{(e)} + \sigma_r^{(h)} \right] N_{fc} = \sigma_r N_{fc},$$ (4.18)

$$\Delta\alpha_{fc} = \left[\sigma_a^{(e)} + \sigma_a^{(h)} \right] N_{fc} = \sigma_a N_{fc}, \tag{4.19}$$

where σ_r and σ_a denote the total free carrier (FC) refraction volume and absorption cross section, respectively. For Si at the 1.55 μm wavelength, the following empirical formulas are often used (Soref and Bennett 1987):

$$\Delta n_{fc} = -(8.8 \times 10^{-22} N_e + 8.5 \times 10^{-18} N_h^{0.8}) \approx \sigma_r N_{fc}, \tag{4.20}$$

$$\Delta\alpha_{fc} = 8.5 \times 10^{-18} N_e + 6.0 \times 10^{-18} N_h = \sigma_a N_{fc}, \tag{4.21}$$

where N_e and N_h are in units of cm^{-3} and $\Delta\alpha_{fc}$ is in units of cm^{-1}. The value of the total absorption cross section σ_a in Equation 4.21 is 1.45×10^{-17} cm^2. The total refraction volume σ_r in Equation 4.20 may be determined by considering that at electron densities around 10^{16}–10^{17} cm^{-3}, for which FC effects in silicon become nonnegligible, the contribution of holes to Δn_{fc} is about 5 times that of electrons. One may thus use an effective value of $\sigma_r = -5.3 \times 10^{-21}$ cm^3 for the refraction volume at these carrier densities (Lin et al. 2007).

The free carrier densities generated by TPA are directly proportional to the squared intensity of the absorbed light. Due to diffusion and recombination, the generated carriers have a finite lifetime, so the induced nonlinear effects in the medium also exhibit a finite response time. In the absence of an external applied field, the spatial and temporal dependence of the FC density, $N_{fc}(\mathbf{r}, t)$, generated by TPA is governed by the diffusion equation

$$\frac{\partial N_{fc}}{\partial t} + \frac{N_{fc}}{\tau_{rec}} - D\nabla^2 N_{fc} = \frac{\alpha_2}{2\hbar\omega} I^2, \tag{4.22}$$

where I is the optical intensity, α_2 the TPA coefficient, τ_{rec} the carrier lifetime due to recombination, and D is the diffusion constant. At the 1.55 μm wavelength, the TPA coefficient of Si is in the range of 0.5–0.8 cm/GW (Dinu et al. 2003, Lin et al. 2007). The recombination lifetime τ_{rec} in bulk Si is typically in the range of a few microseconds (Dimitropoulos et al. 2005). If we define a diffusion time constant τ_{diff} such that $D\nabla^2 N_{fc} = -N_{fc}/\tau_{diff}$, then Equation 4.22 can be expressed as

$$\frac{\partial N_{fc}}{\partial t} + \frac{N_{fc}}{\tau_{fc}} = \frac{\alpha_2}{2\hbar\omega} I^2, \tag{4.23}$$

where the effective FC lifetime τ_{fc} is given by

$$\frac{1}{\tau_{fc}} = \frac{1}{\tau_{rec}} + \frac{1}{\tau_{diff}}. \tag{4.24}$$

In a silicon waveguide, the effective FC carrier lifetime τ_{fc} depends on the waveguide geometry, dimensions, and the defect level in the material and is typically in the order of a nanosecond (Dimitropoulos et al. 2005, Turner-Foster et al. 2010).

4.1.3 Wave propagation in a nonlinear optical waveguide

The formalisms for analyzing nonlinear guided-wave propagation have been extensively developed for optical fibers and waveguides (Marcuse 1991, Agrawal 2013). Here we use a similar approach to Agrawal (2013) and Lin et al. (2007) to derive the nonlinear wave equation for the propagation of a pulse in an integrated optical waveguide with an instantaneous intensity-dependent refractive index. Nonlinear effects due to TPA and free carriers will be considered in Section 4.1.4.

We consider a waveguide oriented along the z direction with a cross-sectional linear index distribution $n_0(x, y)$. For simplicity we assume that both the core and cladding materials are isotropic. The core material exhibits an instantaneous intensity-dependent refractive index of the form given by Equation 4.8. We will neglect any nonlinear effect in the cladding region, either because the material nonlinearity is weak or because the waveguide is strongly confined so that the field in the cladding is small. The starting point in our derivation is the nonlinear wave equation[*]

$$\nabla^2 \mathcal{E} - \frac{1}{c^2} \frac{\partial^2 \mathcal{E}}{\partial t^2} = \mu_0 \frac{\partial^2 \mathcal{P}_L}{\partial t^2} + \mu_0 \frac{\partial^2 \mathcal{P}_{NL}}{\partial t^2}, \tag{4.25}$$

where $\mathcal{E}(\mathbf{r}, t)$ represents a linearly polarized electric field, and $\mathcal{P}_L(\mathbf{r}, t)$ and $\mathcal{P}_{NL}(\mathbf{r}, t)$ are the linear and nonlinear polarizations, respectively. Taking the Fourier transform of Equation 4.25, we get

$$\nabla^2 E(\mathbf{r}, \omega) + k^2 E(\mathbf{r}, \omega) = -\mu_0 \omega^2 P_L(\mathbf{r}, \omega) - \mu_0 \omega^2 P_{NL}(\mathbf{r}, \omega), \tag{4.26}$$

---

[*] Equation 4.25 neglects the vectorial nature of the waveguide mode, which could become important for high-index contrast waveguides in certain applications. A full-vectorial treatment must include the term $-\nabla(\nabla \cdot \mathcal{E})$ on the left-hand side of Equation 4.25 to take into account polarization coupling.

where $k = \omega/c$ and $E(\mathbf{r}, \omega)$ is the Fourier transform of the electric field defined as

$$E(\mathbf{r},\omega) = \int_{-\infty}^{\infty} \mathcal{E}(\mathbf{r},t)e^{-j\omega t}dt. \tag{4.27}$$

The Fourier transforms $P_L(\mathbf{r}, \omega)$ and $P_{NL}(\mathbf{r}, \omega)$ of the linear and nonlinear polarizations are similarly defined and can be related to the electric field by

$$P_L(\mathbf{r},\omega) = \varepsilon_0\chi^{(1)}(x,y,\omega)E(\mathbf{r},\omega), \tag{4.28}$$

$$P_{NL}(\mathbf{r},\omega) = \varepsilon_0\varepsilon_{r,NL}(x,y,\omega)E(\mathbf{r},\omega). \tag{4.29}$$

In the above equations, $\chi^{(1)}$ and $\varepsilon_{r,NL}$ are the linear susceptibility and nonlinear relative permittivity, respectively, which are both functions of (x, y) to account for the spatial index distribution of the waveguide cross section. Upon substituting Equations 4.28 and 4.29 into Equation 4.26, we obtain

$$\nabla^2 E(\mathbf{r},\omega) + n_0^2 k^2 E(\mathbf{r},\omega) = -\varepsilon_{r,NL}k^2 E(\mathbf{r},\omega), \tag{4.30}$$

where $n_0^2 = 1 + \chi^{(1)}$. Equation 4.30 is the nonlinear wave equation for the electric field in the frequency domain. The term on the right-hand side describes the effect of the nonlinear permittivity on light propagation in the waveguide.

We now solve Equation 4.30 for the case of a pulse signal modulating a carrier frequency ω_0. For a pulse propagating in the z direction with a bandwidth centered around frequency ω_0, we can express the electric field in the waveguide as

$$\mathcal{E}(\mathbf{r},t) = \phi(x,y)\tilde{A}(z,t)e^{j(\omega_0 t - \beta_0 z)}, \tag{4.31}$$

where $\phi(x,y)$ is the waveguide mode and $\tilde{A}(z, t)$ is the slowly varying pulse envelope. The Fourier transform of Equation 4.31 is

$$E(\mathbf{r},\omega) = \phi(x,y)A(z,\omega - \omega_0)e^{-j\beta_0 z}, \tag{4.32}$$

where

$$A(z,\omega - \omega_0) = \int_{-\infty}^{\infty} \tilde{A}(z,t)e^{-j(\omega - \omega_0)t}dt. \tag{4.33}$$

The waveguide mode satisfies the eigenvalue equation

$$\left[\nabla_T^2 + n_0^2(x,y,\omega)k^2\right]\phi(x,y) = \beta^2(\omega)\phi(x,y),\tag{4.34}$$

where ∇_T^2 is the transverse Laplacian operator and $\beta(\omega)$ is the linear propagation constant. In Equation 4.31, we define the propagation constant β_0 to be the value of $\beta(\omega)$ at the center frequency, $\beta_0 = \beta(\omega_0)$. By substituting Equation 4.32 into the wave equation (4.30) and performing separation of variables, we obtain

$$\nabla_T^2\phi + (n_0^2 + \varepsilon_{r,NL})k^2\phi = \beta_{NL}^2\phi,\tag{4.35}$$

$$j2\beta_0\frac{\partial A}{\partial z} - (\beta_{NL}^2 - \beta_0^2)A = 0,\tag{4.36}$$

where we have made use of the slowly varying envelope approximation $\partial^2 A/\partial z^2 \ll j\beta_0\partial A/\partial z$ in Equation 4.36. From Equation 4.35 we identify the separation constant β_{NL} as the nonlinear propagation constant of the waveguide in the presence of the nonlinear relative permittivity $\varepsilon_{r,NL}$. The value of β_{NL} can be determined by expressing it as a perturbation of the linear propagation constant β: $\beta_{NL} = \beta(\omega) + \Delta\beta_{NL}$. Making the approximation

$$\beta_{NL}^2 = (\beta + \Delta\beta_{NL})^2 \approx \beta^2 + 2\beta\Delta\beta_{NL},\tag{4.37}$$

and substituting the result into Equation 4.35, we get

$$\varepsilon_{r,NL}k^2\phi(x,y) = 2\beta\Delta\beta_{NL}\phi(x,y).\tag{4.38}$$

Upon multiplying Equation 4.38 by $\phi^*(x,y)$ and integrating over the transverse plane, we obtain

$$\Delta\beta_{NL} = \frac{k^2}{2\beta}\frac{\int_C \varepsilon_{r,NL}\,|\phi(x,y)|^2\,dx\,dy}{\int |\phi(x,y)|^2\,dx\,dy}.\tag{4.39}$$

Note that the integral in the numerator is taken only over the cross section C of the core since we assume that only the waveguide core exhibits nonlinearity.

Next we substitute the solution for $\Delta\beta_{NL}$ into Equation 4.36 to obtain the equation for the pulse envelope $A(z, \omega - \omega_0)$. Using the approximation

$$\beta_{NL}^2 - \beta_0^2 = (\beta_{NL} + \beta_0)(\beta_{NL} - \beta_0) \approx 2\beta_0(\beta + \Delta\beta_{NL} - \beta_0), \tag{4.40}$$

we can simplify Equation 4.36 as follows:

$$\frac{\partial A}{\partial z} + j(\beta - \beta_0)A = -j\Delta\beta_{NL}A. \tag{4.41}$$

The above equation describes the propagation of the pulse envelope in the frequency domain. Chromatic dispersion effects are manifested through the frequency dependence of the propagation constant, $\beta(\omega)$. To account for linear loss in the waveguide, we can add the term $\alpha_0 A/2$, where α_0 is the linear absorption coefficient, to the left-hand side of Equation 4.41:

$$\frac{\partial A}{\partial z} + j[\beta(\omega) - \beta_0]A + \frac{\alpha_0}{2}A = -j\Delta\beta_{NL}A. \tag{4.42}$$

For narrowband pulses, we may assume that the linear loss is approximately constant over the frequency spectrum of the signal.

To convert the nonlinear wave equation (4.42) into the time domain, we expand $\beta(\omega)$ in terms of a Taylor series around ω_0 to get (Agrawal 2013)

$$\frac{\partial A}{\partial z} + j(\Delta\omega\beta_1 + \tfrac{1}{2}\Delta\omega^2\beta_2 + \cdots)A + \frac{\alpha_0}{2}A = -j\Delta\beta_{NL}A, \tag{4.43}$$

where $\Delta\omega = \omega - \omega_0$ and β_n is the nth-order derivative of β: $\beta_n = (d^n\beta/d\omega^n)_{\omega=\omega_0}$. Taking the inverse Fourier transform of the above equation and keeping the first two terms in the Taylor series, we obtain

$$\frac{\partial \tilde{A}}{\partial z} + \beta_1\frac{\partial \tilde{A}}{\partial t} - \frac{j\beta_2}{2}\frac{\partial^2 \tilde{A}}{\partial t^2} + \frac{\alpha_0}{2}\tilde{A} = -j\Delta\tilde{\beta}_{NL}\tilde{A}, \tag{4.44}$$

where $\tilde{A}(z, t)$ is the inverse Fourier transform of $A(z, \omega - \omega_0)$ as defined by Equation 4.33. Equation 4.44 is the nonlinear wave equation for pulse propagation in the time domain. The terms β_1 and β_2 account for the effects of phase velocity dispersion (or group delay) and group velocity dispersion (GVD), respectively. Equation 4.44 is general in that the term $\Delta\tilde{\beta}_{NL}$ can account for any type of nonlinearity in the waveguide core, including free carrier-induced nonlinearity.

For the case where the waveguide core has Kerr nonlinearity, the nonlinear relative permittivity is given by $\varepsilon_{r,NL} = 3\chi^{(3)}|E(\mathbf{r},\omega)|^2$. Substituting this expression into Equation 4.39 and making use of Equation 4.32 for the electric field, we obtain the nonlinear contribution to the propagation constant as follows:

$$\Delta\beta_{NL} = \frac{3\chi^{(3)}k^2}{2\beta} \frac{\int_C |\phi(x,y)|^4 \, dx\,dy}{\int |\phi(x,y)|^2 \, dx\,dy} |A(z,\omega-\omega_0)|^2 . \tag{4.45}$$

Taking the inverse Fourier transform of the above expression, we get

$$\Delta\tilde{\beta}_{NL} = \frac{3\chi^{(3)}k}{2n_{eff}} \frac{\int_C |\phi(x,y)|^4 \, dx\,dy}{\int |\phi(x,y)|^2 \, dx\,dy} |\tilde{A}(z,t)|^2, \tag{4.46}$$

where $n_{eff} = c\beta_0/\omega_0$ is the effective index of the waveguide and we have also made the approximation $\beta \approx n_{eff}k$.

It is convenient to normalize the pulse amplitude $\tilde{A}$ so that $|\tilde{A}|^2$ gives the power in the waveguide. Since the time-averaged power in the waveguide is given by

$$P(z,\omega) = n_{eff}\varepsilon_0 c \int \langle\mathcal{E}^2(\mathbf{r},t)\rangle dx\,dy = 2n_{eff}\varepsilon_0 c |\tilde{A}(z,t)|^2 \int |\phi(x,y)|^2 \, dx\,dy, \tag{4.47}$$

we normalize Equation 4.44 by making the substitution

$$\tilde{A} \to \frac{\tilde{A}}{\left(2n_{eff}\varepsilon_0 c \int |\phi(x,y)|^2 \, dx\,dy\right)^{1/2}} .$$

The resulting equation is

$$\frac{\partial\tilde{A}}{\partial z} + \beta_1 \frac{\partial\tilde{A}}{\partial t} - \frac{j\beta_2}{2}\frac{\partial^2\tilde{A}}{\partial t^2} + \frac{\alpha_0}{2}\tilde{A} = -j\gamma|\tilde{A}|^2\tilde{A}, \tag{4.48}$$

where the nonlinear coefficient γ is given by

$$\gamma = \frac{3\chi^{(3)}\omega}{4n_{\text{eff}}^2\varepsilon_0 c^2 A_{\text{eff}}},\tag{4.49}$$

and A_{eff} is the effective mode area of the waveguide defined as

$$A_{\text{eff}} = \frac{\left(\int |\phi(x,y)|^2\,dx\,dy\right)^2}{\int_C |\phi(x,y)|^4\,dx\,dy}.\tag{4.50}$$

From Equation 4.12 we obtain the relationship between $\chi^{(3)}$ and the nonlinear index n_2 in the core material as $\chi^{(3)} = 4n_c^2 n_2 \varepsilon_0 c/3$, so the nonlinear coefficient γ is also given by[*]

$$\gamma = \left(\frac{n_c}{n_{\text{eff}}}\right)^2 \frac{n_2\omega}{cA_{\text{eff}}}.\tag{4.51}$$

For high-index contrast waveguides, the core index can be substantially larger than the effective index, leading to a factor $(n_c/n_{\text{eff}})^2$ enhancement in the nonlinear coefficient γ compared to that in low-index contrast waveguides. Additional enhancement also comes from the reduced effective mode area A_{eff} of strongly confined waveguides compared to those with weak modal confinement.

Analytical solution of the nonlinear wave equation (4.48) can be obtained for some simple cases. In particular, for narrowband pulses, we can neglect the effects of phase velocity dispersion and GVD and solve the simplified equation

$$\frac{\partial\tilde{A}}{\partial z} + \frac{\alpha_0}{2}\tilde{A} = -j\gamma|\tilde{A}|^2\,\tilde{A}.\tag{4.52}$$

---

[*] In most of the literature, the nonlinear coefficient is given as $\gamma = n_2\omega/cA_{\text{eff}}$, without the factor $(n_c/n_{\text{eff}})^2$. This is a good approximation for low-index contrast waveguides. However, for high-index contrast waveguides where the effective index is significantly smaller than the core index, analytical solutions of the nonlinear wave equation in Equation 4.48 are in better agreement with numerical simulations based on the exact equation in (4.25) if the factor $(n_c/n_{\text{eff}})^2$ is included in the nonlinear coefficient.

Writing $\tilde{A}(z, t) = U(z, t)e^{j\varphi(z,t)}$, where $U(z,t) = |\tilde{A}(z, t)|$, we substitute it into Equation 4.52 and separate the real and imaginary parts to get

$$\frac{\partial U}{\partial z} + \frac{\alpha_0}{2} U = 0, \tag{4.53}$$

$$\frac{\partial \varphi}{\partial z} = -\gamma U^2(z,t). \tag{4.54}$$

From Equation 4.53 we obtain the solution for the pulse amplitude,

$$U(z,t) = U(0,t)e^{-\alpha_0 z/2}. \tag{4.55}$$

Substituting the above result into Equation 4.54 and solving for the nonlinear phase φ, we get

$$\varphi(z,t) = \varphi(0,t) - \gamma U^2(0,t)z_{\text{eff}}, \tag{4.56}$$

where z_{eff} is the effective propagation distance given by

$$z_{\text{eff}} = \frac{1 - e^{-\alpha_0 z}}{\alpha_0}. \tag{4.57}$$

The solution for the pulse envelope can thus be expressed as

$$\tilde{A}(z,t) = \tilde{A}(0,t)e^{-\alpha_0 z/2}e^{j\varphi(z,t)}, \tag{4.58}$$

with the nonlinear phase given by

$$\varphi(z,t) = -\gamma |\tilde{A}(0,t)|^2 z_{\text{eff}}. \tag{4.59}$$

Equation 4.58 shows that as the pulse propagates in the nonlinear waveguide, it retains the initial temporal shape of the envelope except for a decrease in the amplitude due to linear loss. On the other hand, the signal acquires a nonlinear phase shift φ, which depends on the initial pulse power profile $|\tilde{A}(0, t)|^2$. In particular, since φ changes with time, the pulse acquires a frequency chirp given by $\delta\omega = \partial\varphi/\partial t$, which leads to spectral broadening of the signal as it propagates along the nonlinear waveguide.

For a monochromatic wave at frequency ω_0, the solution is simply

$$A(z) = A_0 e^{-\alpha_0 z/2} e^{j\varphi(z)}, \tag{4.60}$$

where the nonlinear phase shift is $\varphi(z) = -\gamma |A_0|^2 z_{\text{eff}}$. In a microring resonator, the acquired nonlinear phase due to SPM causes a shift in the resonant frequency. The total nonlinear phase accumulated over one round-trip of the microring is $\phi_{\text{NL}} = -\gamma |A_0|^2 L_{\text{eff}}$, where L_{eff} is the effective circumference of the microring defined as

$$L_{\text{eff}} = \frac{1 - e^{-2\alpha_0 \pi R}}{\alpha_0} = 2\pi R \left(\frac{a_{\text{rt}}^2 - 1}{2 \ln a_{\text{rt}}} \right). \tag{4.61}$$

In the above expression, $a_{\text{rt}} = e^{-\alpha_0 \pi R}$ is the round-trip amplitude attenuation in the microring. To determine the shift in the resonant frequency caused by the nonlinear phase, we note that, for a change Δn_{NL} in the effective index of the microring waveguide, the resonant frequency ω_m of resonance mode m experiences a shift given by $\Delta \omega_{\text{NL}} = -(\omega_m/n_g)\Delta n_{\text{NL}}$, where n_g is the group index. By writing the nonlinear round-trip phase as $\phi_{\text{NL}} = -\Delta n_{\text{NL}}(\omega_m/c)2\pi R = -\Delta n_{\text{NL}}(2m\pi/n_r)$, where n_r is the effective index, we obtain the following expression for the resonant frequency shift:

$$\Delta \omega_{\text{NL}} = \omega_m \left(\frac{n_r}{n_g} \right) \frac{\phi_{\text{NL}}}{2m\pi} = -\frac{\omega_m}{2m\pi} \left(\frac{n_r}{n_g} \right) \gamma |A_0|^2 L_{\text{eff}}. \tag{4.62}$$

The above equation shows that the resonant frequency is red shifted if the nonlinear coefficient γ (or n_2) is positive, which is typically the case for Kerr nonlinearity in most integrated optics materials near the 1.55 μm wavelength. We will study SPM effects in microring resonators in more detail in Section 4.2.

4.1.4 Free carrier-induced nonlinear effects in an optical waveguide

In Section 4.1.2 we gave a brief review of the nonlinear effects associated with free carriers generated by TPA in a semiconductor material whose bandgap is less than twice the photon energy ($E_g < 2\hbar\omega$). To incorporate these effects into the nonlinear wave equation, we write the total nonlinear relative permittivity in the waveguide core as

$$\varepsilon_{r,NL} = \varepsilon_{r,K} + \varepsilon_{r,fc}, \tag{4.63}$$

where $\varepsilon_{r,K}$ and $\varepsilon_{r,fc}$ are the contributions from Kerr nonlinearity and free carriers, respectively. TPA manifests itself as an intensity-dependent nonlinear loss, which can be accounted for by adding an imaginary term to the Kerr coefficient, $n_2 - j\alpha_2/2k$, where α_2 is the TPA coefficient and $k = \omega/c$. The nonlinear relative permittivity due to the Kerr effect is thus given by

$$\varepsilon_{r,K} = 3\chi^{(3)} |E(\mathbf{r},\omega)|^2 = 4n_c^2\varepsilon_0 c(n_2 - j\alpha_2/2k)|E(\mathbf{r},\omega)|^2, \tag{4.64}$$

where n_c is the linear refractive index of the waveguide core. The contribution of free carriers to the nonlinear permittivity is given by the FC susceptibility defined in Equation 4.15. Using Equations 4.18 and 4.19, we can write the nonlinear relative permittivity due to free carriers in terms of the refraction volume and absorption cross section as

$$\varepsilon_{r,fc} = \chi_{fc} = 2n_c(\sigma_r - j\sigma_a/2k)N_{fc}, \tag{4.65}$$

where $N_{fc}(\mathbf{r}, t)$ is the FC density in the waveguide core. We can now evaluate the contribution of free carriers to the nonlinear propagation constant of the waveguide by using Equation 4.39,

$$\Delta\tilde{\beta}_{fc} = \frac{k^2}{2\beta} \frac{\int_C \varepsilon_{r,fc} |\phi(x,y)|^2 \, dx\, dy}{\int |\phi(x,y)|^2 \, dx\, dy} = \frac{n_c k}{n_{eff}}(\sigma_r - j\sigma_a/2k)\bar{N}_{fc}, \tag{4.66}$$

where $\bar{N}_{fc}(z,t)$ is the spatially averaged FC density over the cross section of the waveguide core,

$$\bar{N}_{fc}(z,t) = \frac{\int_C N_{fc}(\mathbf{r},t) |\phi(x,y)|^2 \, dx\, dy}{\int |\phi(x,y)|^2 \, dx\, dy}. \tag{4.67}$$

Combining the Kerr and FC-induced nonlinear effects, we can write the nonlinear wave equation as

$$\frac{\partial\tilde{A}}{\partial z} + \beta_1 \frac{\partial\tilde{A}}{\partial t} - \frac{j\beta_2}{2}\frac{\partial^2\tilde{A}}{\partial t^2} + \frac{\alpha_0}{2}\tilde{A} = -j\gamma_K |\tilde{A}|^2 \tilde{A} - j\gamma_{fc}\bar{N}_{fc}\tilde{A}, \tag{4.68}$$

where the nonlinear Kerr and FC coefficients are given by

$$\gamma_K = \frac{n_c^2 k}{n_{eff}^2 A_{eff}}\left(n_2 - \frac{j\alpha_2}{2k}\right), \tag{4.69}$$

$$\gamma_{fc} = \frac{n_c k}{n_{eff}}\left(\sigma_r - \frac{j\sigma_a}{2k}\right). \tag{4.70}$$

The above expressions show that, in an optical waveguide, Kerr nonlinear effects are enhanced by a factor of $(n_c/n_{eff})^2$, whereas FC-induced effects are enhanced by a factor of (n_c/n_{eff}). Also note that since the FC refraction volume σ_r is negative while the Kerr coefficient n_2 is typically positive, the nonlinear phase shifts due to FC and Kerr effects are in opposite directions to each other.

In the absence of an external applied field, the free carriers generated from TPA diffuse throughout the waveguide core according to Equation 4.22. Taking the spatial average of this equation over the waveguide cross section, we can write the result as (Lin et al. 2007)

$$\frac{\partial \bar{N}_{fc}}{\partial t} + \left(\frac{1}{\tau_{rec}} + \frac{1}{\tau_{diff}}\right)\bar{N}_{fc} = \bar{G}, \tag{4.71}$$

where $\bar{G}$ is the spatially averaged FC generation rate and τ_{diff} is the diffusion time constant defined by

$$\frac{\int_C D\nabla^2 N_{fc}\,|\phi(x,y)|^2\,dx\,dy}{\int |\phi(x,y)|^2\,dx\,dy} = -\frac{\bar{N}_{fc}}{\tau_{diff}}. \tag{4.72}$$

The FC generation rate $\bar{G}$ can be computed from the attenuation of the optical intensity due to TPA,

$$\bar{G} = -\frac{1}{2\hbar\omega}\overline{\frac{\partial I}{\partial z}} = -\frac{1}{2\hbar\omega}\frac{\int_C (\partial I/\partial z)|\phi(x,y)|^2\,dx\,dy}{\int |\phi(x,y)|^2\,dx\,dy}. \tag{4.73}$$

Since the optical intensity of the waveguide mode is

$$I(x,y,z) = n_{eff}\varepsilon_0 c\langle\mathcal{E}^2(\mathbf{r},t)\rangle = 2n_{eff}\varepsilon_0 c\,|\phi(x,y)|^2|\tilde{A}(z,t)|^2, \tag{4.74}$$

we have

$$\frac{\partial I}{\partial z} = 2n_{\text{eff}}\varepsilon_0 c \,|\phi(x,y)|^2 \left(2U\frac{\partial U}{\partial z}\right), \tag{4.75}$$

where $U = |\tilde{A}(z,t)|$. The rate of attenuation of the field amplitude due to TPA can be determined from

$$\frac{\partial U}{\partial z} = \text{Im}\{\Delta\tilde{\beta}_K\}U, \tag{4.76}$$

where, according to Equation 4.46,

$$\text{Im}\{\Delta\tilde{\beta}_K\} = -\frac{n_c^2\varepsilon_0 c\alpha_2}{n_{\text{eff}}}\left(\frac{\int_C |\phi(x,y)|^4 \, dx\,dy}{\int |\phi(x,y)|^2 \, dx\,dy}\right)U^2. \tag{4.77}$$

Using the above results in Equation 4.73, we obtain the following expression for the average FC generation rate:

$$\bar{G} = \frac{n_c^2\alpha_2}{2\hbar\omega}(2\varepsilon_0 c)^2\left(\frac{\int_C |\phi(x,y)|^4 \, dx\,dy}{\int |\phi(x,y)|^2 \, dx\,dy}\right)^2 U^4. \tag{4.78}$$

Upon performing power normalization of the pulse envelope, we finally get

$$\bar{G} = \left(\frac{n_c}{n_{\text{eff}}}\right)^2 \frac{\alpha_2 U^4}{2\hbar\omega A_{\text{eff}}^2} = \zeta_{\text{TPA}}U^4, \tag{4.79}$$

where the effective mode area A_{eff} is given in Equation 4.50. The equation governing the time evolution of the average FC density in Equation 4.71 can now be expressed as

$$\frac{\partial \bar{N}_{\text{fc}}}{\partial t} + \frac{\bar{N}_{\text{fc}}}{\tau_{\text{fc}}} = \zeta_{\text{TPA}}\,|\tilde{A}(z,t)|^4, \tag{4.80}$$

where τ_{fc} is the effective FC lifetime defined by

$$\frac{1}{\tau_{\text{fc}}} = \frac{1}{\tau_{\text{rec}}} + \frac{1}{\tau_{\text{diff}}}. \tag{4.81}$$

In general, the effects of FCA and FCD become important only for optical pulses with high peak powers or large pulse widths compared to the carrier lifetime. To quantify the relative magnitudes of FCA and TPA, Lin et al. (2007) considered the propagation of a Gaussian pulse with initial power profile $|A(0,t)|^2 = P_0 \exp(-t^2/T_0^2)$ and determined the ratio of the maximum absorption due to FCA (α_{max}^{FCA}) to that due to TPA (α_{max}^{TPA}). The result can be expressed as

$$r_a = \frac{\alpha_{max}^{FCA}}{\alpha_{max}^{TPA}} = \frac{n_c}{n_{eff}} \frac{\sigma_a \mathcal{E}_p}{2\sqrt{2}\hbar\omega A_{eff}}, \tag{4.82}$$

where $\mathcal{E}_p = \sqrt{\pi}P_0 T_0$ is the energy of the Gaussian pulse. The above expression shows that r_a depends only on the pulse energy. FCA effects can be neglected if $r_a \ll 1$, which is typically satisfied for pulse energy $\mathcal{E}_p < 30\,\text{pJ}$ in a Si waveguide around the 1.55 μm wavelength. A similar ratio can also be defined to assess the relative magnitude of FC-induced index change to that due to Kerr nonlinearity. Since an important manifestation of the nonlinear index change is spectral broadening due to frequency chirping, one can look at the growth rate of the frequency chirp with respect to the propagating distance,

$$C = \frac{\partial(\delta\omega)}{\partial z} = \frac{\partial}{\partial z}\left(\frac{\partial\varphi}{\partial t}\right). \tag{4.83}$$

The ratio of the maximum chirp growth due to FCD (C_{max}^{FC}) to that due to Kerr nonlinearity (C_{max}^{Kerr}) is found to be (Lin et al. 2007)

$$r_C = \frac{C_{max}^{FC}}{C_{max}^{Kerr}} = \frac{n_c}{n_{eff}} \frac{\alpha_2 |\sigma_r| \mathcal{E}_p}{2\sqrt{\pi}\hbar\omega n_2 A_{eff}}, \tag{4.84}$$

which also depends only on the pulse energy. The effects of FC-induced index change can be neglected if $r_C \ll 1$, which is typically satisfied for pulse energy $\mathcal{E}_p < 10\,\text{pJ}$ in a Si waveguide around 1.55 μm.

For the general case where both the effects of Kerr nonlinearity and free carriers are nonnegligible, solution of the coupled differential equations in Equations 4.68 and 4.80 can be obtained using a numerical method such as the finite difference method. When the effect of FCA is small ($\gamma_{fc}'' = \text{Im}\{\gamma_{fc}\} \approx 0$), a semi-analytical solution can also be obtained if phase velocity dispersion and GVD are neglected ($\beta_1 = 0$, $\beta_2 = 0$). Writing the solution for the pulse envelope as $\tilde{A}(z, t) = U(z, t)e^{j\varphi(z,t)}$ and substituting it into Equation 4.68, we get

$$\frac{\partial U}{\partial z} + \frac{\alpha_0}{2} U = -\gamma_K'' U^3, \tag{4.85}$$

$$\frac{\partial \varphi}{\partial z} = -\gamma_K' U^2 - \gamma_{fc}' \bar{N}_{fc}, \tag{4.86}$$

where γ_K' and γ_K'' are the real and imaginary parts of the Kerr nonlinear coefficient and γ_{fc}' is the real part of the FC coefficient. We solve Equation 4.85 first to obtain the following expression for the pulse power, $P(z, t) = U^2(z, t)$:

$$P(z,t) = \frac{P_{in}(t)e^{-\alpha_0 z}}{1 + 2\gamma_K'' z_{eff} P_{in}(t)}, \tag{4.87}$$

where z_{eff} is the effective propagation distance given in Equation 4.57 and $P_{in}(t) = P(0, t)$ is the initial power profile of the pulse. Next we substitute $\tilde{A}(z, t) = U(z, t)e^{j\varphi(z,t)}$ into Equation 4.80 to get the equation for the FC density,

$$\frac{d\bar{N}_{fc}}{dt} + \frac{\bar{N}_{fc}}{\tau_{fc}} = \zeta_{TPA}P^2(z,t). \tag{4.88}$$

The above equation can be solved to give

$$\bar{N}_{fc}(z,t) = \zeta_{TPA}e^{-t/\tau_{fc}} \int_{-\infty}^{t} P^2(z,t')e^{t'/\tau_{fc}} dt' = \zeta_{TPA}Q(z,t)e^{-2\alpha_0 z}e^{-t/\tau_{fc}}, \tag{4.89}$$

where

$$Q(z,t) = \int_{-\infty}^{t} \frac{P_{in}^2(t')e^{t'/\tau_{fc}} dt'}{[1 + 2\gamma_K'' z_{eff} P_{in}(t')]^2}. \tag{4.90}$$

The integral in Equation 4.90 can be numerically computed for a given initial pulse power profile $P_{in}(t)$. Substituting Equation 4.89 for the FC density $\bar{N}_{fc}$ into Equation 4.86, we can now solve for the nonlinear phase shift φ. The result can be expressed as $\varphi = \varphi_K + \varphi_{fc}$, where φ_K and φ_{fc} are the nonlinear phase shifts due to Kerr and FC nonlinearity, respectively:

$$\varphi_K(z,t) = -\gamma_K' \int_0^z P(z',t)dz',$$ (4.91)

$$\varphi_{fc}(z,t) = -\gamma_{fc}' \zeta_{TPA} e^{-t/\tau_{fc}} \int_0^z Q(z',t)e^{-2\alpha_0 z'} dz'.$$ (4.92)

The effects of FCA and FCD on the propagation of a Gaussian pulse at the 1.55 μm wavelength in a Si waveguide are illustrated in Figure 4.1. The waveguide has a silicon core ($n_c = 3.45$) with cross-sectional dimensions 450×250 nm² surrounded by SiO₂ cladding. The effective index of the fundamental TE mode is $n_{eff} = 2.5$ and the effective mode area is $A_{eff} = 0.15$ μm². The linear propagation loss in the waveguide is assumed to be 2.5 dB/cm. The Kerr and FC parameters for Si at 1.55 μm are $n_2 = 4.5 \times 10^{-14}$ cm²/W, $\alpha_2 = 0.8$ cm/GW, $\sigma_r = -5.3 \times 10^{-21}$ cm³, and $\sigma_a = 1.45 \times 10^{-17}$ cm². We assume the carrier lifetime τ_{fc} in the Si waveguide to be 1 ns. The Gaussian pulse has an initial power profile given by $P_{in}(t) = P_0 \exp(-t^2/T_0^2)$.

Figures 4.1a–c show the pulse power ($P(t) = |A(z,t)|^2$), nonlinear phase ($\varphi(t)$), and FC carrier density ($\bar{N}_{fc}(t)$) in the waveguide after a propagation distance of 1 mm. The results are obtained for a fixed peak input power $P_0 = 1$ W but different pulse widths T_0 of 0.01 ns, 0.1 ns, and 1 ns, which correspond to pulse energies of 1.8 pJ, 18 pJ, and 180 pJ, respectively. For the 1.8 pJ pulse, the ratios r_a and r_C given by Equations 4.82 and 4.84 are calculated to be 0.065 and 0.34, respectively, indicating that TPA and Kerr-induced index change dominate over FC effects. In particular, we observe that the nonlinear phase is negative during the leading edge of the pulse before becoming positive as the FCD effect becomes more prominent. For higher pulse energies, both nonlinear absorption and index change are dominated by FC effects, with the nonlinear phase and FC density exhibiting long tails due to the large carrier lifetime. Figures 4.1d–f show the effects of increased peak pulse power P_0 for a fixed pulse width T_0 of 0.1 ns. The values of P_0 are 1 W, 2 W, and 5 W, corresponding to pulse energies of 0.18 nJ, 0.36 nJ, and 0.9 nJ, respectively. At high pulse energies, strong FCA causes the pulse shape to become severely attenuated and distorted, as evident in Figure 4.1d.

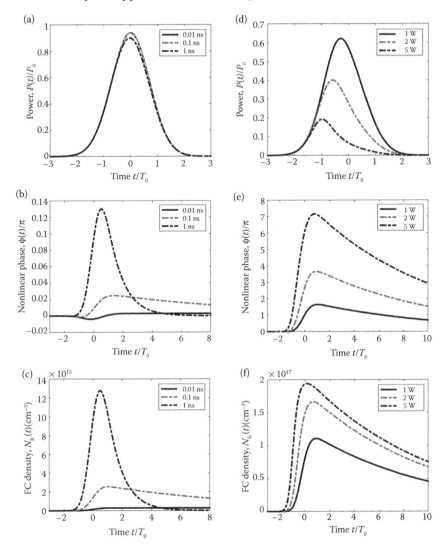

Figure 4.1 Power, nonlinear phase shift, and FC density in a Si waveguide after 1 mm propagation distance for: (a)–(c) fixed peak input pulse power $P_0 = 0.1$ W and pulse width $T_0 = 0.01$ ns, 0.1 ns, and 1 ns, (d)–(f) fixed input pulse width $T_0 = 0.1$ ns and peak input pulse power $P_0 = 1$ W, 2 W, and 5 W.

4.1.5 FWM in a nonlinear optical waveguide

When two or more light waves at different frequencies propagate in a nonlinear optical waveguide, a variety of processes such as XPM, parametric amplification, and frequency conversion can occur. In this section we consider the process of four wave mixing

(FWM), which involves the interaction of four monochromatic waves at frequencies ω_1, ω_2, ω_3, and ω_4 in an optical waveguide with intensity-dependent refractive index. In a typical FWM scheme, two waves called the pump beams at frequencies ω_1 and ω_2 and a signal wave at frequency ω_3 are launched into the waveguide. FWM then gives rise to an idler wave at frequency ω_4. Energy conservation requires that the generated frequency satisfy the relation $\omega_4 = \omega_1 + \omega_2 - \omega_3$.

To analyze the FWM process, we assume that the waveguide is oriented along the z direction with linear refractive index distribution over its cross section given by $n_0(x, y)$. Both the core and cladding materials are isotropic, with the core material possessing Kerr nonlinearity. The two pump beams and the signal wave co-propagate in the waveguide in the positive z direction and are assumed to have the same linear polarization. Consequently, the generated idler wave will also have the same linear polarization and propagate in the same direction. We express the electric field of the ith wave at frequency ω_i in the waveguide as

$$\mathcal{E}_i(\mathbf{r}, t) = E_i(\mathbf{r})e^{j\omega_i t} + \text{c.c..} \tag{4.93}$$

Due to nonlinear interactions of the four waves, the total nonlinear polarization in the waveguide will contain components at various combinations of the four frequencies. However, only those components at frequency ω_i will drive the electric field of the ith wave. Writing the nonlinear polarization at frequency ω_i as

$$\mathcal{R}_{NL}(\mathbf{r}, t, \omega_i) = P_{NL}(\mathbf{r}, \omega_i)e^{j\omega_i t} + \text{c.c.,} \tag{4.94}$$

and substituting the expressions in Equations 4.93 and 4.94 into the wave equation (4.25), we obtain the equation for the field amplitude $E_i(\mathbf{r})$,

$$\nabla^2 E_i(\mathbf{r}) + n_0^2 k_i^2 E(\mathbf{r}) = -\mu_0 \omega_i^2 P_{NL}(\mathbf{r}, \omega_i), \tag{4.95}$$

where $k_i = \omega_i/c$. Next, we write the electric field $E_i(\mathbf{r})$ in the waveguide in terms of the waveguide mode $\phi_i(x, y)$ and amplitude $A_i(z)$,

$$E_i(\mathbf{r}) = \phi_i(x, y)A_i(z)e^{-j\beta_i z}, \tag{4.96}$$

where $\beta_i = \beta(\omega_i)$ is the propagation constant at frequency ω_i. The transverse modal distribution $\phi_i(x, y)$ satisfies the eigenvalue equation

$$\nabla_T^2 \phi_i(x,y) + n_0^2 k_i^2 = \beta_i^2 \phi_i(x,y), \tag{4.97}$$

where ∇_T^2 denotes the transverse Laplacian operator. By substituting Equation 4.96 into Equation 4.95, we get

$$-j2\beta_i\phi_i(x,y)\frac{dA_i}{dz}e^{-j\beta_i z} = -\mu_0\omega_i^2 P_{NL}(\mathbf{r},\omega_i), \tag{4.98}$$

where we have made use of Equation 4.97 and the slowly varying amplitude approximation $d^2A_i/dz^2 \ll j\beta_i dA_i/dz$. Upon multiplying Equation 4.98 by $\phi_i^*(x,y)$ and integrating over the transverse plane, we obtain the equation for the amplitude of the ith beam,

$$\frac{dA_i}{dz} = -\frac{j\mu_0\omega_i^2}{2\beta_i}\frac{\left\langle P_{NL}(\mathbf{r},\omega_i),\phi_i^*\right\rangle}{\left\langle |\phi_i|^2\right\rangle}e^{j\beta_i z}, \tag{4.99}$$

where the angled brackets denote integration over the x–y plane.

In a medium with Kerr nonlinearity, the total nonlinear polarization is given by

$$\mathscr{R}_{NL}(\mathbf{r},t) = \varepsilon_0\chi^{(3)}\mathscr{E}(\mathbf{r},t)\mathscr{E}(\mathbf{r},t)\mathscr{E}(\mathbf{r},t), \tag{4.100}$$

where $\mathscr{E} = \mathscr{E}_1 + \mathscr{E}_2 + \mathscr{E}_3 + \mathscr{E}_4$ is the total electric field. Writing each of the fields $\mathscr{E}_i$ in the form of Equation (4.93) and substituting the total field into Equation 4.100, we obtain the following expressions for the nonlinear polarization at each frequency ω_i:

$$P_{NL}(\mathbf{r},\omega_1) = 3\varepsilon_0\chi^{(3)}\left[|E_1|^2\,E_1 + 2\left(|E_2|^2 + |E_3|^2 + |E_4|^2\right)E_1 + 2E_2^*E_3E_4\right],$$
$$\tag{4.101}$$

$$P_{NL}(\mathbf{r},\omega_2) = 3\varepsilon_0\chi^{(3)}\left[|E_2|^2\,E_2 + 2\left(|E_1|^2 + |E_3|^2 + |E_4|^2\right)E_2 + 2E_1^*E_3E_4\right],$$
$$\tag{4.102}$$

$$P_{NL}(\mathbf{r},\omega_3) = 3\varepsilon_0\chi^{(3)}\Big[|E_3|^2 E_3 + 2\big(|E_1|^2 + |E_2|^2 + |E_4|^2\big)E_3 + 2E_1 E_2 E_4^*\Big],$$

(4.103)

$$P_{NL}(\mathbf{r},\omega_4) = 3\varepsilon_0\chi^{(3)}\Big[|E_4|^2 E_4 + 2\big(|E_1|^2 + |E_2|^2 + |E_3|^2\big)E_4 + 2E_1 E_2 E_3^*\Big].$$

(4.104)

By substituting the above expressions into Equation 4.99, we obtain the equations for the amplitudes $A_i(z)$ of the waves. For example, to obtain the equation for the pump beam at frequency ω_1, we substitute Equation 4.101 into Equation 4.99 to get

$$\frac{dA_1}{dz} = -\frac{j3\chi^{(3)}k_1^2}{2\beta_1\langle|\phi_1|^2\rangle}\Bigg[\langle|\phi_1|^4\rangle|A_1|^2\,A_1 + 2\sum_{k=2}^{4}\langle|\phi_1|^2|\phi_k|^2\rangle|A_k|^2\,A_1$$

$$+ 2\langle\phi_1^*\phi_2^*\phi_3\phi_4\rangle A_2^* A_3 A_4 e^{-j\Delta\beta z}\Bigg],$$

(4.105)

where $\Delta\beta = (\beta_3 + \beta_4) - (\beta_1 + \beta_2)$ is the wave vector mismatch. Performing the power normalization

$$A_i \rightarrow \frac{A_i}{\big(2n_i\varepsilon_0 c\langle|\phi_i|^2\rangle\big)^{1/2}},$$

where $n_i = n_{\mathrm{eff}}(\omega_i)$ is the effective index of mode i, we can express Equation 4.105 as (Agrawal 2013)

$$\frac{dA_1}{dz} = -\frac{j3\chi^{(3)}k_1}{4\varepsilon_0 c}\Big[f_{11}|A_1|^2\,A_1 + 2\big(f_{12}|A_2|^2 + f_{13}|A_3|^2 + f_{14}|A_4|^2\big)A_1$$

$$+ 2f_{1234}A_2^* A_3 A_4 e^{-j\Delta\beta z}\Big],$$

(4.106)

where

$$f_{1k} = \frac{1}{n_1 n_k}\frac{\langle|\phi_1|^2|\phi_k|^2\rangle}{\langle|\phi_1|^2\rangle\langle|\phi_k|^2\rangle},$$

(4.107)

$$f_{1234} = \frac{\left\langle \phi_1^* \phi_2^* \phi_3 \phi_4 \right\rangle}{\prod\limits_{k=1}^{4} \left[n_k \left\langle |\phi_k|^2 \right\rangle \right]^{1/2}}.$$

(4.108)

If the four frequencies are spaced not too far apart, we can approximate the waveguide modes as being nearly identical, $\phi_i \approx \phi$ and $n_i \approx n_{eff}$, in which case Equation 4.106 simplifies to

$$\frac{dA_1}{dz} = -j\gamma(\omega_1) \Big[|A_1|^2 \, A_1 + 2 \big(|A_2|^2 + |A_3|^2 + |A_4|^2 \big) A_1$$
$$+ 2A_2^* A_3 A_4 e^{-j\Delta\beta z} \Big],$$

(4.109)

where

$$\gamma(\omega) = \frac{3\chi^{(3)}\omega}{4n_{eff}^2 \varepsilon_0 c^2 A_{eff}} = \left(\frac{n_c}{n_{eff}} \right)^2 \frac{n_2 \omega}{c A_{eff}},$$

(4.110)

$$A_{eff} = \frac{\left\langle |\phi|^2 \right\rangle^2}{\left\langle |\phi|^4 \right\rangle}.$$

(4.111)

The rightmost expression in Equation 4.110 is obtained using the relation $\chi^{(3)} = 4n_c^2 n_2 \varepsilon_0 c/3$, which is found from Equation 4.12. We also note that the above expressions for the nonlinear coefficient and effective mode area are the same as those in Equations 4.50 and 4.51.

In a similar manner, we can also derive the equations for A_2, A_3, and A_4 to get

$$\frac{dA_2}{dz} = -j\gamma(\omega_2) \Big[|A_2|^2 A_2 + 2 \big(|A_1|^2 + |A_3|^2 + |A_4|^2 \big) A_2 + 2A_1^* A_3 A_4 e^{-j\Delta\beta z} \Big],$$

(4.112)

$$\frac{dA_3}{dz} = -j\gamma(\omega_3) \Big[|A_3|^2 A_3 + 2 \big(|A_1|^2 + |A_2|^2 + |A_4|^2 \big) A_3 + 2A_1 A_2 A_4^* e^{j\Delta\beta z} \Big],$$

(4.113)

$$\frac{dA_4}{dz} = -j\gamma(\omega_4) \Big[|A_4|^2 A_4 + 2 \big(|A_1|^2 + |A_2|^2 + |A_3|^2 \big) A_4 + 2A_1 A_2 A_3^* e^{j\Delta\beta z} \Big].$$

(4.114)

In the above equations, the first term in the square brackets on the right-hand side is responsible for SPM, while the second term gives rise to XPM effects. The third term is responsible for the mutual interactions between the four waves, which leads to the generation and amplification of the new frequency ω_4. We can also include linear propagation loss in the waveguide by adding the term $\alpha_0 A_i/2$ to the left-hand side of Equations 4.109 and 4.112 through 4.114.

Solution of the nonlinear coupled wave equations (4.109) and (4.112) through (4.114) in general requires a numerical approach. For the simplified case where linear loss can be neglected and there is no depletion of the pump beams ($|A_1|^2$, $|A_2|^2 \gg |A_3|^2$, $|A_4|^2$), analytical solutions can be obtained. Neglecting XPM effects induced by the signal and idler waves, the equations for the pump beams reduce to

$$\frac{dA_1}{dz} = -j\gamma \left(|A_1|^2 + 2|A_2|^2 \right) A_1, \tag{4.115}$$

$$\frac{dA_2}{dz} = -j\gamma \left(|A_2|^2 + 2|A_1|^2 \right) A_2, \tag{4.116}$$

where we have also made the simplification that $\gamma(\omega_1) \approx \gamma(\omega_2)$. The above equations can be solved to give

$$A_1(z) = \sqrt{P_1}\, e^{-j\gamma(P_1 + 2P_2)z}, \tag{4.117}$$

$$A_2(z) = \sqrt{P_2}\, e^{-j\gamma(P_2 + 2P_1)z}, \tag{4.118}$$

where $P_1 = |A_1(0)|^2$ and $P_2 = |A_2(0)|^2$ are the initial powers of the pump waves. Substituting the above solutions into the equations for the signal and idler waves and neglecting the SPM terms, we obtain

$$\frac{dA_3}{dz} = -j2\gamma \left[2P_{avg} A_3 + P_0 A_4^* e^{j(\Delta\beta - 6\gamma P_{avg})z} \right], \tag{4.119}$$

$$\frac{dA_4^*}{dz} = j2\gamma \left[2P_{avg} A_4^* + P_0 A_3 e^{-j(\Delta\beta - 6\gamma P_{avg})z} \right], \tag{4.120}$$

where $P_{avg} = (P_1 + P_2)/2$ and $P_0 = \sqrt{P_1 P_2}$ are the arithmetic and geometric averages, respectively, of the pump powers. Solutions of Equations 4.119 and 4.120 are given by

$$A_3(z) = (a_3 e^{gz} + b_3 e^{-gz}) e^{-j\Omega z}, \tag{4.121}$$

$$A_4(z) = (a_4 e^{gz} + b_4 e^{-gz}) e^{-j\Omega z}, \tag{4.122}$$

where

$$g = \sqrt{(2\gamma P_0)^2 - (\gamma P_{avg} + \Delta\beta/2)^2}, \tag{4.123}$$

$$\Omega = 3\gamma P_{avg} - \Delta\beta/2. \tag{4.124}$$

The coefficients a_3, b_3, a_4, and b_4 in Equations 4.121 and 4.122 can be determined from the boundary conditions of the signal and idler waves. The parameter g is called the parametric gain of the FWM process. From Equation 4.123 we see that gain is achieved for a range of values of the wave vector mismatch $\Delta\beta$ such that the term under the square root sign is positive, that is, when $(2\gamma P_0)^2 > (\gamma P_{avg} + \Delta\beta/2)^2$.

Of practical interest is the case of degenerate FWM where the two pump beams A_1 and A_2 are identical. Relabeling the pump, signal, and idler waves as A_p, A_s, A_i, and their frequencies as ω_p, ω_s, ω_i, respectively, we obtain the equations for the three waves as follows:

$$\frac{dA_p}{dz} = -j\gamma(\omega_p)\left[|A_p|^2 A_p + 2(|A_s|^2 + |A_i|^2)A_p + 2A_p^* A_s A_i e^{-j\Delta\beta z}\right], \tag{4.125}$$

$$\frac{dA_s}{dz} = -j\gamma(\omega_s)\left[|A_s|^2 A_s + 2(|A_p|^2 + |A_i|^2)A_s + A_p^2 A_i^* e^{j\Delta\beta z}\right], \tag{4.126}$$

$$\frac{dA_i}{dz} = -j\gamma(\omega_i)\left[|A_i|^2 A_i + 2(|A_p|^2 + |A_s|^2)A_i + A_p^2 A_s^* e^{j\Delta\beta z}\right], \tag{4.127}$$

where $\Delta\beta = \beta_s + \beta_i - 2\beta_p$. The frequency of the idler wave is given by $\omega_i = 2\omega_p - \omega_s$. Under the assumption of no pump depletion, Equations 4.125 through 4.127 simplify to

$$\frac{dA_p}{dz} = -j\gamma |A_p|^2 A_p, \tag{4.128}$$

$$\frac{dA_s}{dz} = -j\gamma \left(2|A_p|^2 A_s + A_p^2 A_i^* e^{j\Delta\beta z}\right), \tag{4.129}$$

$$\frac{dA_i}{dz} = -j\gamma \left(2|A_p|^2 A_i + A_p^2 A_s^* e^{j\Delta\beta z}\right), \tag{4.130}$$

where we have also made the approximation $\gamma(\omega_p) \approx \gamma(\omega_s) \approx \gamma(\omega_i) = \gamma$. The solution of the pump wave is

$$A_p(z) = \sqrt{P_p} e^{-j\gamma P_p z}, \tag{4.131}$$

where $P_p = |A_p(0)|^2$ is the input pump power. The solutions for the signal and idler waves can be expressed as

$$A_s(z) = [a_s \cosh(gz) + b_s \sinh(gz)]e^{-j\Omega z}, \tag{4.132}$$

$$A_i(z) = [a_i \cosh(gz) + b_i \sinh(gz)]e^{-j\Omega z}, \tag{4.133}$$

where

$$g = \sqrt{(\gamma P_p)^2 - (K/2)^2}, \tag{4.134}$$

$$\Omega = 2\gamma P_p - K/2, \tag{4.135}$$

$$K = \Delta\beta + 2\gamma P_p. \tag{4.136}$$

From Equations 4.134 and 4.136, we find that gain is achieved for wave vector mismatch values in the range $-4\gamma P_p < \Delta\beta < 0$. The maximum achievable gain is $g_{max} = \gamma P_p$, which occurs when $K = 0$ or $\Delta\beta = -2\gamma P_p$.

If we assume that only the signal wave and pump wave are present at the input of the waveguide, then we have the boundary conditions $|A_i(0)|^2 = 0$ and $|A_s(0)|^2 = P_s$, where P_s is the initial power of the signal wave. In addition, we also have

$$\frac{dA_s}{dz}\bigg|_{z=0} = -j2P_p A_s(0), \tag{4.137}$$

which can be obtained by evaluating Equation 4.129 at $z = 0$. Applying the above boundary conditions to the solutions in Equations 4.132 and 4.133 and making use of the power conservation requirement $|A_s(z)|^2 + |A_i(z)|^2 = P_s$, we obtain

$$A_s(z) = \sqrt{P_s}[\cosh(gz) - (jK/2g)\sinh(gz)]e^{-j\Omega z}, \tag{4.138}$$

$$A_i(z) = j\sqrt{P_s}\left(\frac{\gamma P_p}{g}\right)\sinh(gz)e^{-j\Omega z}. \tag{4.139}$$

The above solutions describe a power conversion process in which the idler wave grows exponentially with the propagating distance while the signal wave is being depleted. The power conversion efficiency after the waves travel a length L of the waveguide is defined as

$$\eta = \frac{|A_i(L)|^2}{P_s} = \left(\frac{\gamma P_p}{g}\right)^2 \sinh^2(gL). \tag{4.140}$$

Another useful parameter characterizing the FWM process is the coherence length, which is defined as $L_c = 2\pi/|\Delta\beta|$. The coherence length gives the maximum waveguide length over which power conversion can occur in the presence of wave vector mismatch $\Delta\beta$ (Agrawal 2013).

4.2 Optical Bistability and Instability in a Nonlinear Microring Resonator

When an input monochromatic light wave is tuned to a resonant frequency of a nonlinear microring resonator, the light intensity in the microring is greatly amplified due to resonance, which can lead to enhanced nonlinear effects. In a microring resonator with intensity-dependent refractive index, the simplest manifestation of nonlinearity is SPM, which causes an intensity-dependent shift in the resonant frequencies of the microring. At high powers, multiple solutions of the field inside the resonator can exist for a single input power value. Some of these solutions may be unstable, which give rise to nonlinear processes such as optical bistability, self-oscillations, and even chaos. In addition to the important fundamental physics that they entail, these nonlinear processes are also of practical interest for implementing optical switches, threshold-ers, flip-flops, and memory elements.

In this section we study the behavior of a microring resonator with nonlinear refractive index under continuous-wave (CW) excitation. We begin in Section 4.2.1 with the analysis of SPM in an APMR with an intensity-dependent refractive index. Section 4.2.2 shows how optical bistability and instability in the microring can arise due to instantaneous Kerr nonlinearity at moderate to high powers. The influence of free carriers and the effects due to the finite response time of FC-induced nonlinearity will be discussed in Section 4.2.3.

In general, the analysis of the nonlinear dynamics in a microring resonator can be carried out in either the framework of energy coupling or power coupling. However, since the energy coupling formalism is valid only over a limited frequency range around a resonance, it is inadequate for studying the nonlinear behavior of the resonator at large frequency detunings and high powers. An important example of the limitations of the energy coupling formalism is that it fails to predict Ikeda instability (Ikeda et al. 1980), which is caused by the generation and mixing of adjacent cavity modes (Silberberg and Bar-Joseph 1984). In Sections 4.2.1 and 4.2.2 below, we will adopt the power coupling formalism to analyze SPM in microring resonators with instantaneous nonlinearity. On the other hand, since the energy coupling formalism provides a more straightforward approach for incorporating nonlinearity with finite response time, we will use it to obtain an approximate analysis of the resonator dynamics in the presence of free carriers in Section 4.2.3.

4.2.1 SPM and enhanced nonlinearity in a microring resonator

We consider an APMR with radius R and field coupling coefficient κ as shown in Figure 4.2a. The microring waveguide is assumed to have effective index n_r, linear loss coefficient α_0, and instantaneous intensity-dependent refractive index in the waveguide core with Kerr coefficient n_2. The input port of the APMR is excited

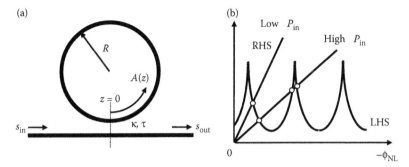

Figure 4.2 (a) Schematic of an APMR for nonlinear dynamic analysis and (b) graphical solutions of Equation 4.145 showing a single real solution for low input power and multiple real solutions for high input power. The LHS curve is the left-hand side expression of Equation 4.145 and the straight lines are the right-hand side (RHS) expression.

by a CW signal s_{in} with frequency ω and power $P_{in} = |s_{in}|^2$. We denote the field circulating inside the microring by $A(z)$, where z is the coordinate along the microring circumference with the origin ($z = 0$) located just after the coupling junction, as indicated in Figure 4.2a. Near a microring resonance, the frequency dispersion due to resonance is much larger than the dispersion of the microring waveguide, so we can neglect the latter and write the equation for the nonlinear propagation of the field inside the microring as

$$\frac{dA}{dz} + \frac{\alpha_0}{2} A = -j\gamma |A|^2 A, \tag{4.141}$$

where γ is the nonlinear coefficient defined in Equation 4.51. The solution of the above equation is given by Equation 4.60,

$$A(z) = A(0)e^{-\alpha_0 z/2}e^{-j\gamma P_0 z_{eff}}, \tag{4.142}$$

where $P_0 = |A(0)|^2$ is the power in the microring at $z = 0$, and $z_{eff} = (1 - e^{-\alpha_0 z})/\alpha_0$ is the effective propagation distance. At the coupling junction between the microring and the waveguide, the following boundary condition must be satisfied:

$$A(0) = \tau A(L)e^{j\phi_L} - j\kappa\sqrt{P_{in}}, \tag{4.143}$$

where $\tau = \sqrt{1-\kappa^2}$ is the transmission coefficient of the coupling junction and $\phi_L = -n_r(\omega/c)2\pi R$ is the linear round-trip phase of the microring. Applying the above boundary condition to the solution in Equation 4.142, we solve for $A(0)$ to get

$$A(0) = \frac{-j\kappa\sqrt{P_{in}}}{1 - \tau a_{rt}e^{j(\phi_L + \phi_{NL})}}.$$

(4.144)

In the above expression, $a_{rt} = e^{-\alpha_0 \pi R}$ is the linear round-trip amplitude attenuation, $\phi_{NL} = -\gamma P_0 L_{eff}$ is the nonlinear round-trip phase, and $L_{eff} = (1 - e^{-2\alpha_0 \pi R})/\alpha_0$ is the effective length of the microring circumference. Taking the absolute square of Equation 4.144, we obtain an equation for ϕ_{NL} which can be expressed as

$$\frac{1}{1 + F \sin^2((\phi_L + \phi_{NL})/2)} = -\frac{\phi_{NL}}{\gamma P_{r,max}L_{eff}},$$

(4.145)

where $F = 4\tau a_{rt}/(1 - \tau a_{rt})^2$ is the contrast factor defined in Equation 2.46, and $P_{r,max}$ is the maximum power in the microring at resonance,

$$P_{r,max} = \frac{\kappa^2 P_{in}}{(1 - \tau a_{rt})^2}.$$

(4.146)

Equation 4.145 can be solved using an iterative numerical technique to obtain the nonlinear phase shift ϕ_{NL} and the power in the microring ($P_0 = -\phi_{NL}/\gamma L_{eff}$) for a given input power P_{in} and linear phase detune ϕ_L. Graphically, we can depict the solutions for ϕ_{NL} as the intersections between the resonance curve given by the left-hand side of Equation 4.145 and the straight line given by the right-hand side, as shown in Figure 4.2b. We find that, for low input powers, Equation 4.145 has only one real solution. At high powers, however, multiple real solutions can occur for a given input power and linear phase detune. Some of these solutions turn out to be unstable and we will discuss them in more detail in Section 4.2.2. For low input powers such that the nonlinear phase shift is small, we can obtain an approximate solution to Equation 4.145 by linearizing the sine function around ϕ_L. This results in the cubic polynomial

$$\left\{1 + F[\sin(\phi_L/2) + (\phi_{NL}/2)\cos(\phi_L/2)]^2\right\}\phi_{NL} = -\gamma P_{r,max}L_{eff}.$$

(4.147)

The solutions for ϕ_{NL} are taken as the real roots of the above polynomial.

Neglecting the small nonlinear phase shift ϕ_{NL} at low input powers, Equation 4.144 indicates that the power in the microring resonator is enhanced by approximately the square of the linear field enhancement factor,

$$\frac{P_0}{P_{in}} \approx \left| \frac{-j\kappa}{1 - \tau a_{rt} e^{j\phi_L}} \right|^2 = FE^2. \tag{4.148}$$

As a result, we expect nonlinear effects in the microring to be also enhanced by a factor FE^2 due to resonance. On the other hand, SPM causes the microring resonant frequency to shift by an amount $\Delta\omega_{NL}$ given by (see Equation 4.62)

$$\frac{\Delta\omega_{NL}}{\omega} = -\left(\frac{n_r}{n_g} \right) \frac{\phi_{NL}}{\phi_L} \approx -\frac{n_r}{n_g} \frac{\gamma P_0}{\beta_r}, \tag{4.149}$$

where $\beta_r = n_r(\omega/c)$ is the linear propagation constant in the microring. Equation 4.149 indicates that the amount of self-detuning is proportional to the power in the microring. Thus, in order to achieve maximum enhancement of nonlinear effects in the resonator, we must initially detune the frequency ω of the input signal from the linear resonance by $\Delta\omega_{NL}$.

Figure 4.3 illustrates the effects of SPM in a silicon APMR. The microring has radius $R = 20\,\mu m$ and coupling coefficient $\kappa = 0.1$.

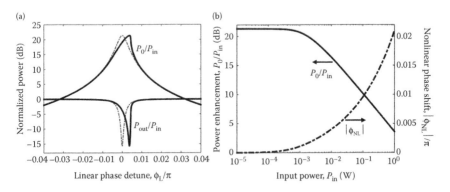

Figure 4.3 Self-phase modulation effects in a nonlinear Si APMR: (a) spectral responses of the output power (P_{out}) and the power in the microring (P_0) normalized to a fixed input power $P_{in} = 5$ mW. Solid lines are responses in the presence of Kerr nonlinearity; dashed lines are responses of the linear device. (b) Power enhancement in the microring (P_0/P_{in}) and self-phase detuning ($|\phi_{NL}|/\pi$) as functions of the input power when the applied CW signal is initially tuned to a microring resonance ($\phi_L = 0$).

The microring waveguide is assumed to have effective index $n_r = 2.5$, linear loss $\alpha_0 = 2.5$ dB/cm, Kerr coefficient $n_2 = 4.5 \times 10^{-14}$ cm^2/W, and effective mode area $A_{eff} = 0.15$ µm^2 near the 1.55 µm wavelength. FCD and FCA are neglected. Figure 4.3a plots the power in the microring (P_0) and the power at the output port (P_{out}) of the APMR as functions of the linear phase detune ϕ_L for a fixed input power $P_{in} = 5$ mW. Also shown for comparison are the spectral responses of the power in the microring and the output power of a linear APMR. The shift in the resonance of the microring due to SPM is apparent from the plot. In addition, we observe that SPM causes the shapes of the spectral responses of the nonlinear APMR to skew to the right (i.e., toward lower frequencies), resulting in a steepening of the right edge of the spectra. Further increase in the input power will lead to bistable behavior, which will be discussed in the next section. Figure 4.3b shows the power enhancement in the microring (P_0/P_{in}) and the nonlinear phase shift ($|\phi_{NL}|$) as functions of the input power when the applied CW signal is initially tuned to a microring resonance ($\phi_L = 0$). We observe that as the input power is increased, the resonant frequency detuning due to SPM also increases, which in turn leads to a reduction in the power enhancement in the microring.

4.2.2   Bistability and self-pulsation in a microring resonator

We saw from the plot in Figure 4.2b that at moderate to high input powers, Equation 4.145 admits multiple real solutions of the nonlinear round-trip phase for a given input power and linear phase detune. Some of these solutions may not be stable and, depending on the nature of the instability, can lead to bistable behavior or oscillation of the field inside the microring. To analyze the nonlinear dynamics in the APMR near an unstable solution, we rewrite the boundary condition (4.143) at the coupling junction of the microring in a slightly different way:

$$A(0, t + T_{rt}) = \tau a_{rt} A(0, t) e^{j\phi_L} e^{-j\xi|A(0,t)|^2} - j\kappa \sqrt{P_{in}}, \qquad (4.150)$$

where $A(0, t)$ is the field in the microring at $z = 0$ and time t, and $A(0, t + T_{rt})$ is the field at the same point but delayed by one round-trip time T_{rt}. For convenience, we have also defined a new nonlinear parameter, $\xi = \gamma L_{eff}$. Equation 4.150 describes the time evolution of the field A in the microring in discrete intervals of T_{rt}. In the theory of nonlinear dynamics, such an equation is called an iterated map. In particular, Equation 4.150 is a 2D map since A is a complex quantity. For a given set of parameters, the iterated map generates a sequence of field values $A_n = A(0, t + nT_{rt})$, which may converge to a fixed value, oscillate about it, or completely diverge from it. The fixed points (or stationary points) of the iterated map are determined by setting $A(0, t + T_{rt}) = A(0, t) = A_0$ in Equation 4.150 and solving for A_0. The result is

$$A_0 = \frac{-j\kappa\sqrt{P_{in}}}{1 - \tau a_{rt}e^{j\theta_L}e^{-j\xi|A_0|^2}}, \tag{4.151}$$

which is the same as Equation 4.144. The fixed-point power in the microring, $P_0 = |A_0|^2$, can thus be determined by solving Equation 4.145. The phase of A_0 is then evaluated by substituting $|A_0|^2 = P_0$ into Equation 4.151.

To determine the dynamic behavior of the field around a fixed-point solution A_0, we perform linear stability analysis by adding a small perturbation $\varepsilon(t)$ to the solution,

$$A(t) = A_0 + \varepsilon(t). \tag{4.152}$$

Substituting the above expression for $A(t)$ into Equation 4.150 and making the approximation

$$e^{-j\xi|A_0+\varepsilon(t)|^2} \approx e^{-j\xi|A_0|^2}\left\{1 - j\xi[A_0^*\varepsilon(t) + A_0\varepsilon^*(t)]\right\},$$

which is valid for small perturbation $\varepsilon(t)$, we get

$$A_0 + \varepsilon(t + T_{rt}) = \tau a_{rt}e^{j\theta_L}[A_0 + \varepsilon(t)]\left\{1 - j\xi\left[A_0^*\varepsilon(t) + A_0\varepsilon^*(t)\right]\right\}e^{-j\xi|A_0|^2}$$
$$-j\kappa\sqrt{P_{in}}. \tag{4.153}$$

Keeping only linear terms in $\varepsilon(t)$ and making use of Equation 4.151, we can further simplify the above expression to

$$\varepsilon(t + T_{rt}) = G\left[\left(1 - j\xi|A_0|^2\right)\varepsilon(t) - j\xi A_0^2\varepsilon^*(t)\right], \tag{4.154}$$

where $G = \tau a_{rt} e^{j\phi_L} e^{-j\xi |A_0|^2}$ is the round-trip phase and attenuation factor. Equation 4.154 and its complex conjugate can be put in matrix form as

$$
\begin{bmatrix} \varepsilon(t+T_{rt}) \\ \varepsilon^*(t+T_{rt}) \end{bmatrix} = \begin{bmatrix} G(1-j\xi |A_0|^2) & -j\xi GA_0^2 \\ j\xi G^*(A_0^*)^2 & G^*(1+j\xi |A_0|^2) \end{bmatrix} \begin{bmatrix} \varepsilon(t) \\ \varepsilon^*(t) \end{bmatrix} \equiv \mathbf{M} \begin{bmatrix} \varepsilon(t) \\ \varepsilon^*(t) \end{bmatrix}.
$$

(4.155)

The above equation describes the discrete time evolution of the perturbation $\varepsilon(t)$. Writing the solution as

$$
\begin{bmatrix} \varepsilon(t+T_{rt}) \\ \varepsilon^*(t+T_{rt}) \end{bmatrix} = e^{\sigma T_{rt}} \begin{bmatrix} \varepsilon(t) \\ \varepsilon^*(t) \end{bmatrix},
$$

(4.156)

and substituting it into Equation 4.155, we get

$$
\begin{bmatrix} G(1-j\xi |A_0|^2)-\lambda & -j\xi GA_0^2 \\ j\xi G^*(A_0^*)^2 & G^*(1+j\xi |A_0|^2)-\lambda \end{bmatrix} \begin{bmatrix} \varepsilon(t) \\ \varepsilon^*(t) \end{bmatrix} = 0,
$$

(4.157)

where $\lambda = e^{\sigma T_{rt}}$ is identified as an eigenvalue of the matrix $\mathbf{M}$. The solution for the perturbation ε after n round-trips is given by

$$
\varepsilon(t+nT_{rt}) = \lambda^n \varepsilon(t).
$$

(4.158)

It is apparent from the above result that the behavior of the perturbation depends on the eigenvalues λ. If both eigenvalues have magnitude less than 1, the perturbation will eventually die off and the perturbed solution $A(t)$ converges to the fixed-point value A_0. In this case we obtain a stable field inside the microring. On the other hand, if either eigenvalue of $\mathbf{M}$ has magnitude greater than 1, the solution is unstable. The nature of the instability depends on the phase φ of the eigenvalue, $\lambda = |\lambda| e^{j\varphi}$. Specifically, if the eigenvalue is real and greater than 1 ($\varphi = 0$ and $|\lambda| > 1$), the perturbed solution $A(t)$ will be repelled from the fixed-point value A_0, to be attracted to nearby fixed points. This situation corresponds to optical bistability. If the eigenvalue is real and less than −1 ($\varphi = \pi$ and $|\lambda| > 1$), the perturbation changes sign every round-trip time T_{rt} according to Equation 4.158. The field in this case oscillates around the fixed point with a period exactly equal to twice the microring round-trip time. This is called period-doubling oscillation or Ikeda instability

(Ikeda et al. 1980). For other values of φ (λ is complex with $|\lambda| > 1$), the solution oscillates with a period equal to $2\pi T_{rt}/\varphi$. This behavior is referred to as self-pulsation.

For the matrix **M** in Equation 4.155, the eigenvalues can be evaluated to give

$$\lambda_\pm = \tau a_{rt} \left(\theta \pm \sqrt{\theta^2 - 1} \right), \tag{4.159}$$

$$\theta = \cos(\phi_L - \xi P_0) + \xi P_0 \sin(\phi_L - \xi P_0). \tag{4.160}$$

The parameter θ depends on the linear phase detune ϕ_L and the power P_0 in the microring. If $\theta^2 < 1$, the eigenvalues are complex with magnitude $|\lambda_\pm| = \tau a_{rt} < 1$, so the solution is stable. If $\theta^2 \geq 1$, the eigenvalues are real. In this case bistability occurs if $\theta > (\tau a_{rt} + 1/\tau a_{rt})/2$ whereas Ikeda instability occurs if $\theta < -(\tau a_{rt} + 1/\tau a_{rt})/2$. Since the eigenvalues are never complex with magnitude greater than 1, self-oscillation other than the Ikeda type cannot occur in a microring with instantaneous Kerr nonlinearity. We will see in Section 4.2.3, however, that if the nonlinearity has a finite response time, such as due to Debye relaxation or free carrier diffusion and recombination, self-pulsation can occur with oscillation periods in the order of the relaxation time (Ikeda et al. 1982, Silberberg and Bar-Joseph 1984).

Ikeda instability typically occurs at very high input powers and large linear phase detunes. It is characterized by self-oscillations at a period exactly equal to twice the microring round-trip time ($2T_{rt}$), and this period is insensitive to changes in the phase detune or input power as long as the condition $\theta < -(\tau a_{rt} + 1/\tau a_{rt})/2$ is met. Thus, there exist certain ranges of ϕ_L and P_{in} values for which period-doubling oscillations can occur. The origin of period-doubling oscillations can be explained in terms of the spontaneous generation of the two cavity modes adjacent to the input frequency due to FWM (Silberberg and Bar-Joseph 1984). Suppose an input CW wave tuned to a frequency ω_0 about half way between two resonance modes ω_m and ω_{m+1} of the microring, as shown in Figure 4.4. As the input power is increased, the cavity modes are shifted to lower frequencies due to SPM (assuming $n_2 > 0$). When the input frequency falls exactly half way between the two modes, spontaneous FWM generates two new frequencies at the cavity modes, which then beat with the input signal to give rise to oscillations at a period equal to 2/FSR, or twice the microring round-trip time.

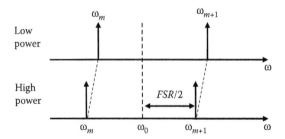

Figure 4.4 Diagram depicting the origin of period-doubling oscillations due to spontaneous FWM generation of the cavity modes ω_m and ω_{m+1} adjacent to the input frequency ω_0. (After Silberberg, Y., Bar-Joseph, I., 1984, *J. Opt. Soc. Am. B*, 1, (4): 662–670.)

As a numerical example illustrating SPM effects in a microring, we show in Figure 4.5 the stability curves of a silicon APMR for two sets of coupling coefficient and linear phase detune values: (a) $\kappa = 0.1$ and $\phi_L = 0.025\pi$, and (b) $\kappa = 0.7$ and $\phi_L = 0.5\pi$. The radius of the microring is $R = 100\ \mu m$. The microring waveguide has effective index $n_r = 2.5$, linear loss $\alpha_0 = 2.5$ dB/cm, Kerr coefficient $n_2 = 4.5 \times 10^{-14}\ cm^2/W$ and effective mode area $A_{eff} = 0.15\ \mu m^2$ near the 1.55 μm wavelength. Both FCD and FCA are neglected. The plots show the fixed-point solutions of the power in the microring (P_0) as a function of the input power (P_{in}). The regions of stability (S), bistability (BS) and Ikeda oscillations (IK) are indicated on the plots. We see that bistability can occur at moderate input powers and small linear phase detunes. On the other hand, the threshold power required to reach Ikeda instability is significantly higher and is typically beyond the power levels that can be handled by most integrated optics waveguides. For this reason, Ikeda instability has not been observed in a microring resonator.

4.2.3 Free carrier effects and self-pulsation in a microring resonator

In the previous section, we saw that in a microring resonator with instantaneous Kerr nonlinearity, period-doubling oscillations can occur at very high powers and large linear phase detunes. In this section, we show that if the nonlinearity has a finite response time, self-pulsation can occur at much lower powers. An example of where this behavior may be observed is in a semiconductor microring resonator with a significant concentration of free carriers generated

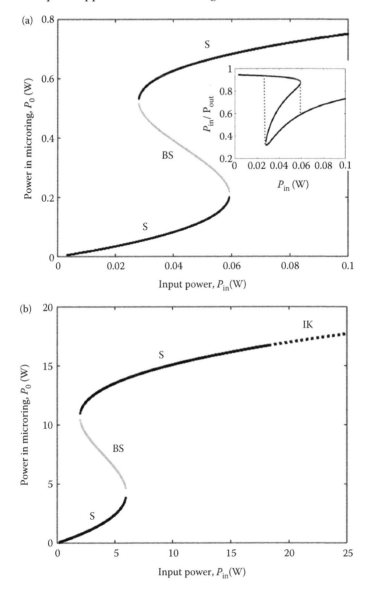

Figure 4.5 Plots of the power in a Si APMR (P_0) versus input power: (a) $\kappa = 0.1$ and $\phi_L = 0.025\pi$, (b) $\kappa = 0.7$ and $\phi_L = 0.5\pi$. Regions of stability, bistability, and Ikeda instability are indicated by S, BS, and IK, respectively. Inset of (a) shows a plot of the power transmission of the APMR (P_{out}/P_{in}) versus the input power.

from TPA. We will find that in such a resonator, the period of oscillation depends on the FC lifetime and that there is an upper limit to this value for which self-pulsation can occur (Armaroli et al. 2011, Malaguti et al. 2011).

Analysis of the nonlinear dynamics of a microring resonator with FC-induced nonlinearity is complicated by the addition of a differential equation governing the evolution of the carrier density. Since we are mainly interested in the behavior of the resonator in the vicinity of a resonance (i.e., small linear phase detunes), it is more expedient to employ the energy coupling formalism for our analysis (Malaguti et al. 2011, Chen et al. 2012). In this model, the nonlinear dynamics of the resonator are described by a system of continuous-time differential equations rather than the discrete-time iterated map in Equation 4.150. It should be kept in mind, however, that the energy coupling formalism only provides an approximate model of the resonator. We will discuss the limitations of this model in predicting the nonlinear dynamic behavior of a microring resonator.

We consider again the APMR in Figure 4.2a, which is excited by an input monochromatic wave at frequency ω with power P_{in}. Let $a(t)$ represent the amplitude of the wave in the microring, which is normalized so that $|a(t)|^2$ gives the total energy in the resonator. The rate of energy coupling between the straight waveguide and the microring is denoted by μ, which is related to the field coupling coefficient κ by $\mu = \kappa/T_{rt}$, where $T_{rt} = 2\pi R/v_g$ is the microring round-trip time and v_g is the group velocity. We assume that the microring waveguide possesses instantaneous Kerr nonlinearity as well as FC-induced nonlinearity. Including both the effects of nonlinear dispersion and nonlinear absorption, the equation governing the evolution of the energy amplitude in the microring can be expressed as

$$\frac{da}{dt} = -j(\Delta\omega - \Delta\omega_{NL})a - (\gamma_L + \gamma_{NL})a - j\mu\sqrt{P_{in}} , \qquad (4.161)$$

where $\Delta\omega = \omega - \omega_0$ is the detuning from the resonant frequency ω_0 of the microring, $\Delta\omega_{NL}$ is the nonlinear frequency shift, and γ_L and γ_{NL} are the linear and nonlinear decay rates, respectively. The nonlinear frequency shift of the microring resonance is related to the nonlinear index change Δn_{NL} by

$$\Delta\omega_{NL} = -\omega_0 \frac{\Delta n_{NL}}{n_g} , \qquad (4.162)$$

where n_g is the group index of the microring waveguide. We can determine the nonlinear index change by summing the contributions from Kerr nonlinearity and FCD,

$$\Delta n_{\mathrm{NL}} = \Delta n_{\mathrm{NL,K}} + \Delta n_{\mathrm{NL,fc}} = \left(\frac{n_c}{n_r}\right)^2 \frac{n_2}{A_{\mathrm{eff}}} \frac{|a(t)|^2}{T_{\mathrm{rt}}} + \left(\frac{n_c}{n_r}\right)\sigma_r N(t), \quad (4.163)$$

where n_c is the refractive index of the waveguide core, n_r the effective index, n_2 the Kerr coefficient, A_{eff} the effective mode area, $N(t)$ the FC density and σ_r is the FC refraction volume. The FC density in the microring is given by Equation 4.80,

$$\frac{dN}{dt} + \frac{N}{\tau_{\mathrm{fc}}} = \left(\frac{n_c}{n_r}\right)^2 \frac{\alpha_2}{2\hbar\omega A_{\mathrm{eff}}^2} \frac{|a(t)|^4}{T_{\mathrm{rt}}^2}, \quad (4.164)$$

where α_2 is the TPA absorption coefficient. In evaluating the Kerr nonlinear index change in Equation 4.163, we have approximated the power in the microring as $|a(t)|^2/T_{\mathrm{rt}}$.

The decay rates γ_{L} and γ_{NL} in Equation 4.161 are related to the linear and nonlinear loss, respectively, in the microring. Specifically, the linear decay rate can be computed from the linear loss coefficient α_0 of the microring waveguide and the coupling coefficient μ as

$$\gamma_{\mathrm{L}} = \frac{\alpha_0 v_g}{2} + \frac{\mu^2}{2}. \quad (4.165)$$

The nonlinear decay rate is due to TPA and FCA and can be evaluated using the nonlinear absorption coefficients in Equations 4.69 and 4.70,

$$\gamma_{\mathrm{NL}} = \frac{\alpha_{\mathrm{NL,K}} v_g}{2} + \frac{\alpha_{\mathrm{NL,fc}} v_g}{2} = \frac{v_g}{2}\left[\left(\frac{n_c}{n_r}\right)^2 \frac{\alpha_2}{A_{\mathrm{eff}}} \frac{|a(t)|^2}{T_{\mathrm{rt}}} + \left(\frac{n_c}{n_r}\right)\sigma_a N(t)\right],$$

$$(4.166)$$

where σ_a is the FCA cross section. Substituting the above results for the nonlinear frequency shift $\Delta\omega_{\mathrm{NL}}$ and nonlinear decay rate γ_{NL} into Equation 4.161, we can summarize the equations for the energy amplitude and FC density in the microring as follows:

$$\frac{da}{dt} = -(j\Delta\omega + \gamma_{\mathrm{L}})a - j\eta_{\mathrm{K}}|a|^2 a - j\eta_{\mathrm{fc}}Na - j\mu\sqrt{P_{\mathrm{in}}}, \quad (4.167)$$

$$\frac{dN}{dt} + \frac{N}{\tau_{fc}} = \varsigma |a|^4,$$

(4.168)

where

$$\eta_K = \eta_K' - j\eta_K'' = \left(\frac{n_c}{n_r}\right)^2 \left(n_2 k_0 - \frac{j\alpha_2}{2}\right) \frac{v_g}{A_{eff} T_{rt}},$$

$$\eta_{fc} = \eta_{fc}' - j\eta_{fc}'' = \frac{n_c}{n_r}\left(\sigma_r k_0 - \frac{j\sigma_a}{2}\right) v_g,$$

$$\varsigma = \left(\frac{n_c}{n_r}\right)^2 \frac{\alpha_2}{2\hbar\omega A_{eff}^2 T_{rt}^2}.$$

Along with Equation 4.168, Equation 4.167 and its complex conjugate form a 3D nonlinear dynamical system. The fixed points a_0 and N_0 of the system are determined by setting $da/dt = 0$ and $dN/dt = 0$ to get

$$N_0 = \tau_{fc}\varsigma |a_0|^4,$$

(4.169)

$$a_0 = \frac{-j\mu\sqrt{P_{in}}}{j\Delta\omega + j\eta_K |a_0|^2 + j\eta_{fc} N_0 + \gamma_L}.$$

(4.170)

To determine the stability of the fixed points, we add a small perturbation to a_0 and N_0 to get $a(t) = a_0 + \varepsilon(t)$, $N(t) = N_0 + \delta n(t)$. Making these substitutions in Equations 4.167 and 4.168 and keeping only first-order terms in $\varepsilon(t)$ and $\delta n(t)$, we obtain

$$\frac{d\varepsilon}{dt} = -(j\Delta\omega + j2\eta_K |a_0|^2 + j\eta_{fc} N_0 + \gamma_L)\varepsilon - j\eta_K |a_0|^2 \varepsilon^* - j\eta_{fc} a_0 \delta n,$$

(4.171)

$$\frac{d\delta n}{dt} = 2\varsigma |a_0|^4 (a_0^*\varepsilon + a_0\varepsilon^*) - \frac{\delta n}{\tau_{fc}}.$$

(4.172)

The above equations can be put in the matrix form, $d\varepsilon/dt = \mathbf{M}\varepsilon$, where $\varepsilon = [\varepsilon(t), \varepsilon^*(t), \delta n(t)]^T$ and the matrix $\mathbf{M}$ is given by

$$\mathbf{M} = \begin{bmatrix} -\Gamma & -j\eta_K a_0^2 & -j\eta_{fc} a_0 \\ j\eta_K^* (a_0^*)^2 & -\Gamma^* & j\eta_{fc}^* a_0^* \\ 2\varsigma |a_0|^2 a_0^* & 2\varsigma |a_0|^2 a_0 & -1/\tau_{fc} \end{bmatrix},$$

(4.173)

$$\Gamma = j\Delta\omega + j2\eta_K |a_0|^2 + j\eta_{fc} N_0 + \gamma_L.$$

It is apparent that the time evolution of the perturbation vector ε is determined by the eigenvalues of the matrix $\mathbf{M}$. In particular, if all the eigenvalues reside in the left half of the complex plane, the fixed-point solutions a_0 and N_0 are stable. If one of the eigenvalues crosses into the right half of the complex plane, the fixed point becomes unstable. In this case bistability occurs if the eigenvalue is real and positive, and Hopf bifurcation occurs if the eigenvalues become complex with positive real parts. The latter case corresponds to self-pulsation with the period of oscillation equal to $|2\pi/\mathrm{Im}\{\lambda\}|$.

Before considering the behavior of the APMR in the presence of free carriers, we first look at the simple case where only instantaneous Kerr nonlinearity is present ($\varsigma = 0, \eta_K'' = 0, \eta_{fc} = 0$). From Equation 4.170 we obtain the following cubic polynomial for the energy $E_0 = |a_0|^2$ in the microring:

$$\left[(\delta + hE_0)^2 + 1 \right] E_0 = E_{in,c}, \tag{4.174}$$

where $\delta = \Delta\omega/\gamma_L$ is the normalized frequency detune, $h = \eta_K'/\gamma_L$ the nonlinear factor and $E_{in,c} = \mu^2 P_{in}/\gamma_L$ is the energy coupled into the microring. The above polynomial has a single real root for $\delta < \sqrt{3}$. For $\delta > \sqrt{3}$, three real roots exist for microring energies in the range

$$-\frac{2\delta + \sqrt{\delta^2 - 3}}{3h} < E_0 < -\frac{2\delta - \sqrt{\delta^2 - 3}}{3h}. \tag{4.175}$$

To evaluate the stability of these roots, we determine the eigenvalues of the matrix $\mathbf{M}$ in Equation 4.173, which in this case reduces to

$$\mathbf{M} = \begin{bmatrix} -\Gamma & -j\eta_K' a_0^2 \\ j\eta_K' (a_0^*)^2 & -\Gamma^* \end{bmatrix}, \tag{4.176}$$

where $\Gamma = j(\Delta\omega + 2\eta_K' E_0) + \gamma_L$. We find that the normalized eigenvalues, $\lambda_n = \lambda/\gamma_L$, of the above matrix are given by

$$\lambda_n = -1 \pm j\sqrt{(\delta + hE_0)(\delta + 3hE_0)}. \tag{4.177}$$

For microring energies in the range given by Equation 4.175, the eigenvalues are real with one being negative and the other positive. Thus the two limits in Equation 4.175 correspond to the lower and upper energy thresholds for observing bistable behavior. Outside this range of energies, the eigenvalues can be real or complex but the real parts are always negative, so the solutions are always stable. Note that the above analysis fails to predict instability of the Ikeda type. This is due to the fact that the energy coupling model neglects neighboring resonance modes of the microring, whose interactions give rise to period-doubling oscillations as discussed in Section 4.2.2.

We next consider the effects of free carriers on the nonlinear dynamics of the APMR (Armaroli et al. 2011, Malaguti et al. 2011). For simplicity, we will assume that FCD dominates over Kerr nonlinearity so that we can set $\eta_K = 0$. In addition, we will also neglect FCA ($\eta''_{fc} = 0$) in the analysis although it can have a significant impact on the threshold powers for observing instability, as will be discussed in the example at the end of the section. With these simplifications, we find that the fixed-point values of the energy in the microring are given by the roots of the fifth-order polynomial

$$\left[(\delta + h E_0^2)^2 + 1\right] E_0 = E_{in,c},\qquad(4.178)$$

where $\delta = \Delta\omega/\gamma_L$, $E_{in,c} = \mu^2 P_{in}/\gamma_L$ and the nonlinear factor is given by $h = \eta'_{fc}\tau_{fc}\varsigma/\gamma_L$. The above polynomial has multiple real roots for microring energies in the range $E^-_{0,bs} < E_0 < E^+_{0,bs}$, where

$$E^\pm_{0,bs} = \left(-\frac{3\delta \pm \sqrt{4\delta^2 - 5}}{5h}\right)^{1/2}.\qquad(4.179)$$

The stability of these roots are determined by evaluating the eigenvalues of the matrix $\mathbf{M}$, which in this case simplifies to

$$\mathbf{M} = \begin{bmatrix} -\Gamma & 0 & -j\eta'_{fc}a_0 \\ 0 & -\Gamma^* & j\eta'_{fc}a_0^* \\ 2\varsigma|a_0|^2 a_0^* & 2\varsigma|a_0|^2 a_0 & -1/\tau_{fc} \end{bmatrix},\qquad(4.180)$$

where $\Gamma = j(\Delta\omega + \eta'_{fc}\tau_{fc}\varsigma E_0^2) + \gamma_L$. The normalized eigenvalues, $\lambda_n = \lambda/\gamma_L$, of the above matrix are found to be solutions of a cubic polynomial of the form

$$\lambda_n^3 + a_2\lambda_n^2 + a_1\lambda_n + a_0 = 0,\qquad(4.181)$$

with the coefficients given by

$$a_2 = 2 + 1/\tau_n,$$
$$a_1 = 1 + 2/\tau_n + \delta_{NL}^2,$$
$$a_0 = (1 + \delta_{NL}^2 + 4hE_0^2\delta_{NL}^2)/\tau_n.$$

In the above expressions, $\tau_n = \tau_{fc}\gamma_L$ is the normalized FC lifetime and $\delta_{NL} = \delta + hE_0^2$ is the normalized nonlinear frequency detune. The normalized FC lifetime can also be interpreted as the ratio of the FC lifetime to the linear cavity lifetime, $\tau_n = \tau_{fc}/\tau_L'$, where $\tau_L' = 1/\gamma_L.$* For microring energies in the range $E_{0,bs}^- < E_0 < E_{0,bs}^+$, where $E_{0,bs}^\pm$ are given by Equation 4.179, it can be shown that one of the eigenvalues obtained from Equation 4.181 is real and positive, indicating that bistability will occur in this range. Outside this energy range, the eigenvalues can become complex with positive real part, which signifies self-pulsation behavior. The threshold energies for the onset of self-pulsation are determined by the Hopf bifurcation points, which are defined as the points where a complex eigenvalue crosses the imaginary axis into the right half-plane. These points can be found using the Routh–Hurwitz stability criterion for a cubic characteristic polynomial, $a_1a_2 = a_0$. We find that, in general, there are two Hopf bifurcation points occurring at the energies:

$$E_{0,sp}^\pm = \left[\frac{-(1-\tau_n)\delta \pm \sqrt{\Delta}}{(2-\tau_n)h} \right]^{1/2}, \tag{4.182}$$

$$\Delta = (\delta^2 + 3) - (\tau_n^2 - 2/\tau_n). \tag{4.183}$$

Self-pulsation occurs for microring energies in the range $E_{0,sp}^- < E_0 < E_{0,sp}^+$. If $E_{0,sp}^\pm$ are complex, self-pulsation cannot occur. We can find the condition for which self-pulsation ceases to occur by setting $\Delta = 0$, yielding the equation

$$\tau_{n,c}^3 - (\delta^2 + 3)\tau_{n,c} - 2 = 0. \tag{4.184}$$

For a given frequency detune δ, solution of the above equation gives the critical value $\tau_{n,c}$, which defines the upper limit of the normalized

---

* Here we define the cavity lifetime τ_L' as the time it takes the wave amplitude $a(t)$, not the energy $|a(t)|^2$, in the microring to decay to $1/e$ of its initial value.

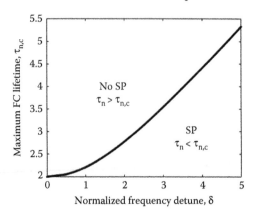

Figure 4.6 Plot of the maximum normalized FC lifetime for which self-pulsation can occur versus the normalized frequency detune. Self-pulsation occurs in the region where $\tau_n < \tau_{n,c}$.

FC lifetime for which self-pulsation can occur. A plot of the critical value $\tau_{n,c}$ versus the normalized frequency detune is shown in Figure 4.6. Self-pulsation can occur only in the region $\tau_n < \tau_{n,c}$. From the plot, we find that in an APMR where FC-induced nonlinearity dominates, self-pulsation is always possible if $\tau_n < 2$.

As a numerical example, we consider a silicon APMR around the 1.55 μm wavelength where FCD is assumed to be the dominant source of nonlinearity. The microring has radius $R = 25$ μm, effective index $n_r = 2.5$, linear loss $\alpha_0 = 2.5$ dB/cm, and field coupling coefficient $\kappa = 0.1$. Assuming a group index $n_g = 4.5$ for the waveguide, we calculate the corresponding energy coupling coefficient to be $\mu = 6.5 \times 10^4$ s$^{-1/2}$ and linear decay rate $\gamma_L = 4$ GHz (or cavity lifetime $\tau_L' = 0.25$ ns). The silicon waveguide has TPA coefficient $\alpha_2 = 0.8$ cm/GW, FC refraction volume $\sigma_r = -5.3 \times 10^{-21}$ cm^3, and effective mode area $A_{eff} = 0.15$ μm^2. We neglect Kerr-induced dispersion and FCA effects. In Figure 4.7a we plot the power in the microring ($P_0 = E_0/T_{rt}$) and the output power of the APMR (P_{out}) as functions of the linear frequency detune for input power $P_{in} = 1$ mW and normalized FC lifetime $\tau_n = 2$. Also shown for comparison are the spectral responses of the device in the absence of any nonlinearity. Since free carriers induce a negative refractive index change in the microring waveguide, the spectral responses of the device are skewed towards positive frequency detunes as the input power is increased. In Figure 4.7b we plot the power in the microring (P_0) versus the input power (P_{in}) for frequency detune $\delta = 2$ and normalized

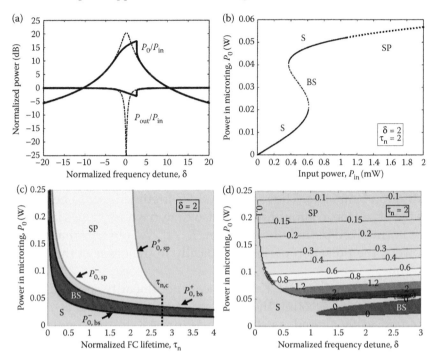

Figure 4.7 (a) Plots of the output power (P_{out}) and power in the microring (P_0) versus the linear frequency detune (δ) for $P_{in} = 1$ mW and $\tau_n = 2$. Solid lines are responses in the presence of TPA and FCD; dashed lines are responses in the linear case. (b) Plot of the power in the microring (P_0) versus the input power (P_{in}) for $\delta = 2.0$ and $\tau_n = 2$, showing regions of stability (S), bistability (BS), and self-pulsation (SP). (c) Threshold powers for bistability and self-pulsation as functions of the normalized FC lifetime τ_n at a fixed $\delta = 2.0$. (d) Stability map of microring power versus frequency detune showing regions of stability, bistability, and self-pulsation for a APMR with $\tau_n = 2$. Values of the contour lines in the SP region are normalized periods of oscillations, T_{sp}/τ'_L.

FC lifetime $\tau_n = 2$. The regions of stability (S), bistability (BS), and self-pulsation (SP) are indicated on the plot. Compared to the results in Figure 4.5 where FC effects are neglected and instantaneous Kerr nonlinearity dominates, the thresholds for bistability in this case occur at much lower powers since FC-induced nonlinearity is much stronger than the Kerr effect in silicon. In addition, self-pulsation can also be observed at relatively low input powers. Note, however, that we have assumed an FC lifetime τ_{fc} of only $2\tau'_L$, or about 0.5 ns, which is smaller than typically observed in a silicon waveguide. In addition, nonlinear loss due to FCA has also been neglected.

To investigate the influence of the FC lifetime on the nonlinear dynamics of the microring, we plot in Figure 4.7c the threshold powers for bistability $(P_{0,\mathrm{bs}}^{\pm} = E_{0,\mathrm{bs}}^{\pm}/T_{\mathrm{rt}})$ and self-pulsation $(P_{0,\mathrm{sp}}^{\pm} = E_{0,\mathrm{sp}}^{\pm}/T_{\mathrm{rt}})$ as functions of the normalized FC lifetime for a fixed frequency detune $\delta = 2$. We find that for any value of τ_{n}, there always exists a range of powers in which bistability will occur. On the other hand, self-pulsation occurs only for values of τ_{n} up to a critical value of 2.8, or a maximum FC lifetime of about 0.7 ns. Figure 4.7d shows the regions of frequency detunes and microring powers for which the device exhibits stable, bistable, and self-pulsation behaviors. The normalized FC lifetime is set to be $\tau_{\mathrm{n}} = 2$. In the SP region, contour lines are also drawn whose values indicate the normalized period of oscillation, $T_{\mathrm{sp}}/\tau_{\mathrm{L}}'$. We observe that at low powers, the pulsation period is typically in the order of the cavity lifetime.

When the effects of Kerr-induced nonlinearity and FCA are also taken into account, the nonlinear behaviors of the Si APMR are noticeably altered, as shown in Figure 4.8. In the simulations, we assume a Kerr coefficient $n_2 = 4.5 \times 10^{-14}$ cm^2/W and FCA cross section $\sigma_{\mathrm{a}} = 1.45 \times 10^{-17}$ cm^2 for the silicon waveguide. Figure 4.8a shows the stability curve of the APMR for frequency detune $\delta = 2$ and normalized FC lifetime $\tau_{\mathrm{n}} = 2$. For comparison, the stability curve obtained when Kerr nonlinearity and FCA are neglected is also shown in the plot. In general, Kerr-induced index change has negligible effect on the nonlinear dynamics of the microring since it is much smaller than FCD. On the other hand, FCA can significantly increase the cavity loss, so higher powers are required to reach bistable and self-pulsation regimes. Another effect of FCA is that it reduces the critical FC lifetime value $\tau_{\mathrm{n,c}}$ and the range of microring powers for which self-pulsation can occur. This can be seen by comparing the plots in Figures 4.7c and 4.8b, which show the threshold powers for bistability and self-pulsation versus τ_{n} for a fixed $\delta = 2$. Figure 4.8c shows the stability map of microring power versus frequency detune for $\tau_{\mathrm{n}} = 2$. Again, the region of self-pulsation behavior is much reduced compared to the plot in Figure 4.7d where FCA is neglected. These results show that it is more difficult to achieve self-pulsation in an APMR when significant FCA is present.

Free carrier-induced self-pulsation in a microring resonator can be explained in terms of the interactions between the optical resonance and the nonlinear dispersion caused by the TPA-generated

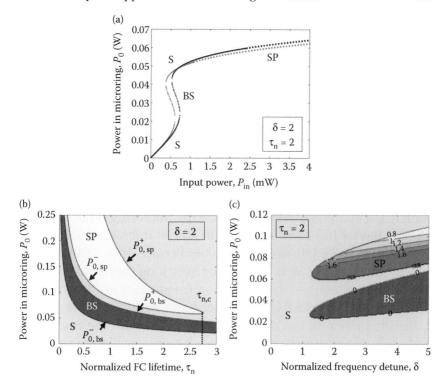

Figure 4.8 (a) Stability plots of a Si APMR showing the power in the microring (P_0) versus the input power (P_{in}) for $\delta = 2$ and $\tau_n = 2$. Black line is the case where Kerr-induced index change and FCA are included; gray line is the case where these effects are neglected. (b) Plot of the threshold powers for bistability and self-pulsation as functions of the normalized FC lifetime τ_n at a fixed $\delta = 2$. (c) Stability map of microring power versus frequency detune showing regions of stability, bistability, and self-pulsation for a microring with $\tau_n = 2$. Values of the contour lines in the SP region are normalized periods of oscillations, T_{sp}/τ_L'.

free carriers. More specifically, near a resonant frequency, the light intensity buildup in the microring causes the FC density to increase, which in turn causes a blue shift in the cavity mode due to FCD. As a result, the light becomes detuned from the resonance, leading to a decrease in the light intensity and hence the FC density in the microring. However, as the FC density decreases, the blue shift in the cavity mode also gets smaller, so the light becomes less detuned from the resonance and begins to build up again in the cavity. The resulting behavior is thus an oscillation of the light intensity in the microring with a period comparable to the cavity lifetime.

Self-pulsation can also arise from the competing interactions between FC-induced index change and the photothermal refraction effect in a semiconductor microring resonator (Zhang et al. 2013). At moderate to high input powers, light absorption in the microring can generate sufficient heat to induce a change in the refractive index of the waveguide via the thermo-optic effect. In a silicon waveguide, the index change due to photothermal refraction is positive, while the FC-induced index change is negative. Thus FCD and photothermal refraction exert a kind of push-and-pull effect on the resonant wavelength of a microring resonator and, if the two processes have comparable time constants, can result in oscillations of the field inside the microring. This self-pulsation behavior has been observed in silicon microring resonators (Priem et al. 2005, Johnson et al. 2006, Pernice et al. 2010, Zhang et al. 2014). However, due to the relatively long thermal time constants in these devices, the oscillation frequencies are typically in the MHz range or lower. On the other hand, self-pulsations caused by the interaction of the nonlinear FCD effect with the resonance mode, which are characterized by much faster oscillation frequencies (in the GHz range), have not been observed, mainly due to the deleterious effect of FCA and the short FC lifetime required to observe this behavior.

4.3   All-Optical Switching in a Nonlinear Microring Resonator

One of the applications of a material with an intensity-dependent refractive index is all-optical switching, which involves the modification of an optical signal either by itself or by another optical signal. The former case, called self-switching, is based on the process of SPM, whereby an optical signal switches itself from a low-power state to a high-power state or vice versa. The latter case typically employs a pump-and-probe switching scheme in which a strong pump beam is used to modulate the amplitude of a weaker probe signal through XPM. In this section, we consider both self-switching and pump-and-probe switching in a nonlinear microring resonator (Van et al. 2002a). In both schemes, a strong input pump signal is used to cause an intensity-dependent shift of the microring resonance, which translates into a modulation of the transmission of either a probe signal or the pump itself.

We begin in Section 4.3.1 by showing that under CW excitation, the buildup of light intensity in the microring near a resonance can lead to a substantial reduction in the nonlinear switching power compared to a nonresonant device such as a Mach–Zehnder interferometer (MZI). In Section 4.3.2, we will treat the problem of nonlinear pulse propagation in a microring resonator and show how SPM in the resonator can lead to self-switching. Section 4.3.3 extends the analysis to the case of pump-and-probe switching in a microring resonator with intensity-dependent nonlinear refractive index. The effects of FCD and FCA will also be examined for both switching schemes.

4.3.1 Enhanced nonlinear switching in a microring resonator

In a typical pump-and-probe switching configuration, a strong pump signal is used to induce a nonlinear phase shift in a weaker probe signal. This phase modulation can be converted into a change in the transmitted amplitude of the probe signal through the use of an interferometric device such as a MZI or a cavity such as a microring resonator. A comparison of the switching efficiency of these two devices—one resonant and the other nonresonant—will serve to highlight the advantage of microring resonators for nonlinear optics applications. For simplicity, we will restrict our analysis below to structures with instantaneous intensity-dependent refractive index under CW excitation.

We consider first a simple switch based on an unbalanced MZI with phase difference $\delta = \phi_2 - \phi_1$ between the two arms as shown in Figure 4.9a. For a CW probe signal at frequency ω_s applied to port 1, its power transmission at port 4 is given by

$$T = \left| \frac{s_{\text{out}}}{s_{\text{in}}} \right|^2 = \cos^2\left(\frac{\delta}{2} \right). \tag{4.185}$$

Suppose an intense CW pump signal at frequency ω_p is injected into the upper arm of the MZI. The pump signal causes a nonlinear phase shift in the probe signal in the upper arm equal to $\phi_{\text{NL}} = -2\gamma P_p k_s L$, where γ is the nonlinear coefficient of the waveguide given in Equation 4.51, P_p is the power of the pump beam, $k_s = \omega_s/c$, and L is the arm length of the MZI. The factor 2 appears because the phase shift due to XPM is twice as large as that due to SPM. The total phase difference of the two MZI arms is thus

$$\delta = \phi_2 - (\phi_1 - 2\gamma P_p k_s L). \tag{4.186}$$

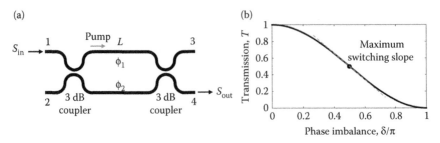

Figure 4.9   (a) Schematic of an unbalanced MZI for pump-and-probe switching and (b) power transmission at port 4 as a function of the total phase imbalance, $\delta = \phi_2 - (\phi_1 + \phi_{NL})$, between the two MZI arms.

Figure 4.9b plots the transmission of the probe signal with respect to the total phase imbalance δ of the MZI. It is clear that the output power of the probe signal, $|s_{out}|^2$, can be modulated by varying the pump power. For example, a high input pump power will cause a large nonlinear phase shift, which leads to a low transmitted probe power. The switching efficiency of the MZI can be estimated from the slope of the transmission curve,

$$\frac{dT}{dP_p} = \frac{dT}{d\delta}\frac{d\delta}{dP_p}. \tag{4.187}$$

The first derivative on the right-hand side ($dT/d\delta$) can be evaluated from Equation 4.185. Its maximum absolute value is 1/2, which occurs when the MZI has a linear phase imbalance of $\phi_2 - \phi_1 = \pi/2$. The second derivative can be computed from Equation 4.186 to give $d\delta/dP_p = 2\gamma k_s L$. The maximum switching slope of the MZI is thus

$$\left.\frac{dT}{dP_p}\right|_{max} = \gamma k_s L, \tag{4.188}$$

which indicates that the switching efficiency increases linearly with the MZI arm length L.

In an APMR the switching efficiency is enhanced by two separate effects. First, the effective nonlinear interaction length L is increased by a factor roughly equal to the resonator finesse, $\mathcal{F} \sim 1/\kappa^2$, where κ is the coupling coefficient. Second, the light intensity inside the microring is amplified by a factor of $FE^2 \sim 1/\kappa^2$ at resonance (see Equation 2.22). Thus we expect a reduction in the switching power proportional to $\sim 1/\kappa^4$ in an APMR compared to an MZI.

To estimate the maximum switching power reduction achievable in a microring resonator, we consider an APMR of radius R, field coupling coefficient κ, and round-trip attenuation a_{rt}. We assume that the APMR is critically coupled so that a steep switching slope can be obtained near a resonance. Under this condition, the linear transmission response of the device is given by Equation 2.45 with $\tau = a_{rt}$,

$$T_{ap} = \left| \frac{S_{out}}{S_{in}} \right|^2 = \frac{2\tau^2(1-\cos\phi)}{1+\tau^4-2\tau^2\cos\phi}, \qquad (4.189)$$

where ϕ is the microring round-trip phase. Suppose a weak CW probe signal tuned near a microring resonance is applied to the input port. The transmission of the probe signal is modulated by applying a strong CW pump with power $P_{in,p}$ to the input port, as shown in Figure 4.10a. The pump beam is assumed to be of the same polarization as the probe but tuned to a different resonant frequency of the microring. Accounting for the nonlinear phase shift induced by the pump beam through XPM, the total round-trip phase of the probe signal in the microring is given by

$$\phi = -(n_r + 2\gamma P_{r,p})k_s L, \qquad (4.190)$$

where n_r is the effective index of the waveguide, $L = 2\pi R$ is the microring circumference, and $P_{r,p} = |A_p|^2$ is the pump power circulating in the microring. Figure 4.10b plots the probe power transmission T_{ap} as a function of the round-trip phase detune $\Delta\phi$ from

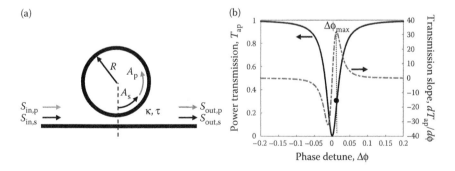

Figure 4.10   (a) Schematic of an APMR for pump-and-probe switching and (b) plots of the power transmission and transmission slope versus phase detune of a critically coupled APMR with $\tau = a_{rt} = 0.99$.

a microring resonance. Application of the pump beam causes the phase detune to decrease (assuming $\gamma > 0$), so that the transmission curve will effectively shift to the right in Figure 4.10b. Thus a probe signal tuned to a point on the right shoulder of the transmission curve will experience a decrease in the transmission.

Figure 4.10b also plots the variation in the slope of the transmission curve, which is given by

$$\frac{dT_{ap}}{d\phi} = \frac{2\tau^2 \kappa^4 \sin\phi}{(1 + \tau^4 - 2\tau^2 \cos\phi)^2}. \tag{4.191}$$

From the plot, we observe that the maximum transmission slope occurs slightly off the resonance. The phase ϕ_{max} at which the maximum slope occurs can be computed from the condition $d^2 T_{ap}/d\phi^2 = 0$. Assuming that the microring has a very high Q factor ($\kappa \ll 1$), we get $\cos\phi_{max} \approx 1 - \kappa^4/6$, which gives the maximum value for the slope of the transmission curve as

$$\left.\frac{dT_{ap}}{d\phi}\right|_{\phi_{max}} \approx \frac{3\sqrt{3}}{8\kappa^2}. \tag{4.192}$$

The switching efficiency of the APMR can be computed from

$$\frac{dT_{ap}}{dP_{in,p}} = \frac{dT_{ap}}{d\phi} \frac{d\phi}{dP_{r,p}} \frac{dP_{r,p}}{dP_{in,p}}, \tag{4.193}$$

where the first derivative on the right-hand side is given by Equation 4.192. The second derivative is evaluated from Equation 4.190 to give $d\phi/dP_{r,p} = -2\gamma k_s L$. The last derivative, $dP_{r,p}/dP_{in}$, gives the pump power enhancement in the microring at ϕ_{max}. From Equation 2.21 we obtain the expression for the pump power in the APMR as follows:

$$P_{r,p} = \frac{\kappa^2 P_{in,p}}{1 + \tau^4 - 2\tau^2 \cos\phi}, \tag{4.194}$$

from which we evaluate the derivative at ϕ_{max} to be $dP_{r,p}/dP_{in,p} \approx 3/4\kappa^2$. Substituting the above results for the derivatives into Equation 4.193, we obtain the maximum switching slope

$$\left.\left|\frac{dT_{ap}}{dP_{in,p}}\right|\right|_{max} \approx \frac{9\sqrt{3}}{8}\left(\frac{\gamma k_s L}{\kappa^4}\right) \propto FE^4, \tag{4.195}$$

where $FE \sim 1/\kappa$ is the field enhancement factor at resonance. The above result shows that compared to an MZI having the same arm length as the microring circumference, the microring resonator has a switching slope that is proportional to the fourth power of the field enhancement factor, or the square of the finesse $\mathcal{F}$ (since $FE^2 = \mathcal{F}/\pi$). For a microring resonator with power coupling efficiency $\kappa^2 \sim 1\%$, the switching slope can be enhanced by as much as four orders of magnitude.

4.3.2 Self-switching of a pulse in a nonlinear microring resonator

In Section 4.2.1 we showed that under CW excitation, SPM effects in a microring resonator with intensity-dependent refractive index are enhanced by approximately the square of the field enhancement factor (FE^2) near a microring resonance. In this section, we extend the analysis to a pulse propagating in a microring resonator with an instantaneous intensity-dependent refractive index. Simulation examples demonstrating self-switching in an add-drop microring resonator will be given. We will also examine how free carrier effects can modify pulse propagation in the microring.

We consider an ADMR with input and output field coupling coefficients κ_1 and κ_2, respectively.[*] The microring is assumed to have radius R, effective index n_r, linear loss coefficient α_0, and nonlinear Kerr coefficient n_2. An input signal of the form $\tilde{s}_{in}(t) = s_{in}(t)e^{j\omega_0 t}$, where $s_{in}(t)$ is the slowly varying pulse envelope, is applied to the input port of the ADMR. The power-normalized wave inside the microring can be expressed as

$$\tilde{A}(z,t) = A(z,t)e^{j(\omega_0 t - \beta_0 z)}, \tag{4.196}$$

where $\beta_0 = n_r(\omega_0)\omega_0/c$. The coordinate z is defined along the circumference of the microring with $z = 0$ located just after the input coupling junction, as shown in Figure 4.2a for an APMR device. In the vicinity of a resonant frequency, the frequency dispersion due to resonance dominates over the dispersion of the microring waveguide. We can thus neglect the effects of both phase velocity dispersion and GVD in Equation 4.48 and write the

---

[*] The analysis in this section is also applicable to an APMR by setting $\kappa_2 = 0$.

equation for the propagation of the pulse envelope $A(z, t)$ inside the microring as

$$\frac{\partial A}{\partial z} + \frac{\alpha_0}{2} A = -j\gamma |A|^2 A, \tag{4.197}$$

where γ is the nonlinear coefficient defined in Equation 4.51. The solution of the above equation is

$$A(z,t) = A(0,t)e^{-\alpha_0 z/2}e^{-j\gamma|A(0,t)|^2 z_{\text{eff}}}, \tag{4.198}$$

where the effective propagation distance z_{eff} is given by Equation 4.57. At the input coupling junction of the ADMR, the field satisfies the boundary condition

$$A(0,t) = \tau_1 A(L,t)e^{j\phi_L} - j\kappa_1 s_{\text{in}}(t), \tag{4.199}$$

where $L = 2\pi R$ is the microring circumference and $\phi_L = -\beta_0 L$ is the linear round-trip phase. At the output coupling junction, the field in the microring is attenuated by a factor $\tau_2 = \sqrt{1-\kappa_2^2}$ due to coupling to the output waveguide. If the coupling is small, the power removed from the microring at the output coupling junction may be lumped into the linear waveguide loss so that the total round-trip amplitude attenuation in the microring can be approximated as $\tau_2 a_{\text{rt}}$, where $a_{\text{rt}} = e^{-\alpha_0 L/2}$. Using Equation 4.198, we can write the solution for $A(L, t)$ as

$$A(L,t) = \tau_2 a_{\text{rt}} A(0,t)e^{j\phi_{\text{NL}}}, \tag{4.200}$$

where $\phi_{\text{NL}} = -\gamma|A(0, t)|^2 L_{\text{eff}}$ and L_{eff} is the effective microring circumference given by

$$L_{\text{eff}} = 2\pi R \left[\frac{(\tau_2 a_{\text{rt}})^2 - 1}{2\ln(\tau_2 a_{\text{rt}})} \right]. \tag{4.201}$$

Upon substituting Equation 4.200 into the boundary condition in Equation 4.199 and solving for $A(0, t)$, we get

$$A(0,t) = \frac{-j\kappa_1 s_{\text{in}}(t)}{1 - \tau_1 \tau_2 a_{\text{rt}} e^{j(\phi_L + \phi_{\text{NL}})}}. \tag{4.202}$$

Taking the absolute square of the above equation and defining the input power $P_{in}(t) = |s_{in}(t)|^2$, we obtain the equation for the nonlinear phase shift ϕ_{NL},

$$\left\{1 + F\sin^2\left[(\phi_L + \phi_{NL})/2\right]\right\}\phi_{NL}(t) = -\gamma L_{eff} FE_{max}^2 P_{in}(t), \qquad (4.203)$$

where $F = 4\tau_1\tau_2 a_{rt}/(1 - \tau_1\tau_2 a_{rt})^2$ is the contrast factor and $FE_{max}^2 = \kappa_1^2/(1 - \tau_1\tau_2 a_{rt})^2$ is the maximum power enhancement in the ADMR. Equation 4.203 can be solved using an iterative numerical technique. Alternatively, for small nonlinear phase shifts, an approximate solution can be obtained by linearizing the sine term around ϕ_L (similar to Equation 4.147). At high input powers, multiple real solutions for ϕ_{NL} may exist which indicate bistable (or multistable) behavior. By substituting the solution for the nonlinear phase shift into Equation 4.202, we obtain the complex envelope $A(0, t)$ of the signal in the microring. Finally, the output signals at the drop port and through port of the ADMR are obtained from

$$s_d(t) = -j\kappa_2\sqrt{a_{rt}}\, A(0, t)e^{j(\phi_L + \phi_{NL})/2}, \qquad (4.204)$$

$$s_t(t) = \tau_1 s_{in}(t) - j\kappa_1\tau_2 a_{rt} A(0, t)e^{j(\phi_L + \phi_{NL})}. \qquad (4.205)$$

In Equation 4.204 we have made the approximation that the nonlinear phase shift at the output coupling junction is equal to half the value of the nonlinear round-trip phase.

Numerical results for a Gaussian pulse at the 1.55 μm wavelength propagating in a nonlinear silicon ADMR are shown in Figure 4.11. We assume that only Kerr nonlinearity is present in the Si waveguide and neglect effects due to TPA and free carriers. The parameters used for the Si waveguide are: effective index $n_r = 2.5$, linear propagation loss $\alpha_0 = 2.5$ dB/cm, Kerr coefficient $n_2 = 4.5 \times 10^{-14}$ cm^2/W, and effective mode area $A_{eff} = 0.15$ μm^2. The microring has radius $R = 5$ μm and equal input and output power coupling coefficients $\kappa^2 = 2\%$. The drop port transmission spectrum of the device has a 3 dB bandwidth of 13 GHz. We apply an input Gaussian pulse having power profile given by $P_{in}(t) = P_0\exp[-(t/T_0)^2]$ with the center frequency tuned to a microring resonance ($\phi_L = 2m\pi$). The pulse width is set to be $T_0 = 50$ ps, corresponding to a signal bandwidth of 5.3 GHz, which is well below the 3 dB bandwidth of the microring so that signal distortion due to the filtering effect of the ADMR is negligible. Figures 4.11a and b show the output powers at the

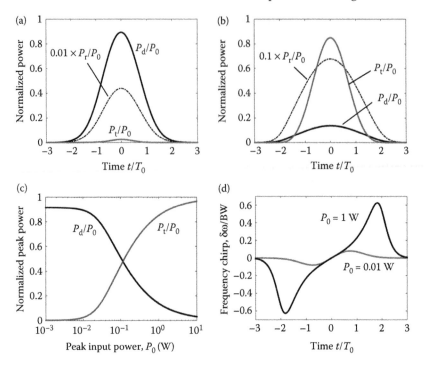

Figure 4.11 Powers at the drop port (P_d) and through port (P_t), and in the microring (P_r) for (a) low peak input power, $P_0 = 0.01$ W, and (b) high peak input power, $P_0 = 1$ W. The pulse width is fixed at $T_0 = 50$ ps. (c) Switching curves of the ADMR showing the dependence of the drop port transmission (P_d/P_0) and through port transmission (P_t/P_0) on the peak input power. (d) Frequency chirp of the signal in the microring for peak input powers P_0 of 0.01 W and 1 W, both with a pulse width of $T_0 = 50$ ps.

drop port (P_d) and through port (P_t) of the microring for peak input pulse power $P_0 = 0.01$ W and 1 W, respectively. These peak power levels correspond to pulse energies of 0.9 pJ and 90 pJ, respectively. The power in the microring (P_r) is also shown for each case. At low input power (Figure 4.11a), the input signal is nearly in resonance with the microring, so most of the pulse power appears at the drop port and very little power appears at the through port. When the input power is increased (Figure 4.11b), the signal becomes detuned with respect to the microring resonance, causing most of the pulse power to be switched to the through port. We note that the peak power in the microring is enhanced by a factor of about 45 for the low input power case and a factor of 7 for the high input power case. This enhancement of the microring power helps amplify the nonlinear effects in the resonator and lower the switching power. Figure 4.11c shows the variations of the peak power transmissions at the drop port (P_d/P_0) and the through port (P_t/P_0) with respect to the input power. From the plot, we find that the maximum switching slope occurs around a peak input power of about 100 mW.

It is also instructive to see how the frequency spectrum of the signal inside the microring is modified by the Kerr nonlinearity. We found in Section 4.1.3 that a pulse propagating in a waveguide with instantaneous Kerr nonlinearity experiences spectral broadening caused by frequency chirping. Due to resonance enhancement of SPM effects, the spectral broadening experienced by a pulse in a microring resonator can be much larger than in a straight waveguide. Figure 4.11d shows the frequency chirp experienced by a Gaussian pulse propagating in the silicon ADMR for two different values of the peak input pulse power: $P_0 = 0.01$ W and 1 W. We see that the frequency spectrum of the signal broadens by around 8% of the bandwidth at the low input power and as much as 60% at the high input power.

The switching speed of a microring resonator with instantaneous nonlinearity is limited by the cavity lifetime. In a semiconductor microring resonator in which substantial TPA also occurs, FCA and dispersion can greatly modify the device response, especially at high input powers and large input pulse widths. In particular, if the FC lifetime is much longer than the cavity lifetime, the switching speed of the device will be mainly limited by the former time constant. To take into account the effects of free carriers on the switching response of an ADMR, the nonlinear wave equation (4.68) must be solved along with Equation 4.80 for a pulse propagating in the microring subject to the boundary condition at the input coupling junction given by Equation 4.199. To simplify the solution, we may model the output coupler of the ADMR as a source of loss and lump it with the linear propagation loss of the microring waveguide, so that the effective linear loss coefficient becomes $\alpha_{0,\text{eff}} = \alpha_0 - \ln(\tau_2)/\pi R$.

The effects of TPA and free carriers on the propagation of a Gaussian pulse in a Si ADMR are illustrated by the plots in Figure 4.12. The microring parameters are the same as those in the previous example. In addition, we also assume the following FC parameters for silicon around the 1.55 μm wavelength: TPA coefficient $\alpha_2 = 0.8$ cm/GW, FC refraction volume $\sigma_r = -5.3 \times 10^{-21}$ cm^3, FCA cross section $\sigma_a = 1.45 \times 10^{-17}$ cm^2, and carrier lifetime $\tau_{fc} = 1$ ns. Figure 4.12a plots the transmitted powers at the drop port (P_d) and

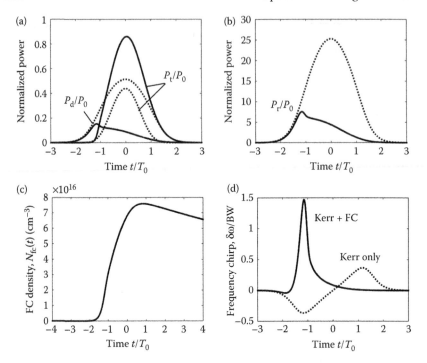

Figure 4.12 Nonlinear response of a Si ADMR to a Gaussian input pulse with peak power $P_0 = 0.1$ W and pulse width $T_0 = 50$ ps: (a) transmitted powers at the drop port (P_d) and through port (P_t); (b) power in the microring (P_r) when both Kerr nonlinearity and FC are present (solid lines) and when only Kerr nonlinearity is present (dotted lines). (c) Time evolution of the FC density (N_{fc}) inside the microring. (d) Frequency chirp of the pulse in the microring when both Kerr and FC nonlinear effects are present (solid line) and when only Kerr nonlinearity is present (dotted line).

through port (P_t) of the ADMR for an input Gaussian pulse with peak power $P_0 = 0.1$ W and pulse width $T_0 = 50$ ps. The pulse power and FC density inside the microring are shown in Figures 4.12b and c, respectively. For comparison, we also show in Figures 4.12a and b the pulse signals (dotted lines) when FC effects are neglected and only Kerr nonlinearity is present in the microring. It is apparent that FC effects significantly distort and attenuate the pulse inside the microring and also the output signal at the drop port. Figure 4.12d compares the frequency chirp of the pulse inside the microring with and without FC effects. In the presence of free carriers, we observe significant positive chirp at the leading edge of the pulse due to a rapid increase in the generated FC density.

4.3.3 Pump-and-probe switching in a nonlinear microring resonator

In a typical all-optical switching application, a control (pump) wave is used to modulate the amplitude of a signal (probe) wave. Such a pump-and-probe switching operation can be accomplished in a microring resonator through the process of XPM, whereby a strong pump wave is used to induce a nonlinear phase shift in the probe signal, causing a modulation of its transmitted amplitude.

To analyze the interaction of the pump and probe waves in a microring resonator with an instantaneous intensity-dependent refractive index, we consider an ADMR with input pulses for both the pump and probe of the form

$$\tilde{s}_{in,1}(t) = s_{in,1}(t)e^{j\omega_1 t}, \quad \tilde{s}_{in,2}(t) = s_{in,2}(t)e^{j\omega_2 t}, \tag{4.206}$$

where $s_{in,1}(t)$ and $s_{in,2}(t)$ are the envelopes of the pump pulse and probe pulse, respectively, and the frequencies ω_1 and ω_2 are tuned to two different resonances of the microring. For simplicity, we assume that the pump and probe waves have the same polarization, and the peak power of the pump pulse is much larger than that of the probe pulse so that we can neglect XPM of the pump by the probe. If we also neglect both phase velocity dispersion and GVD, then the equation governing the pump pulse envelope $A_1(z, t)$ in the microring and its solution are given by Equations 4.197 and 4.198, respectively. The equation for the probe pulse envelope $A_2(z, t)$ in the microring is

$$\frac{\partial A_2}{\partial z} + \frac{\alpha_0}{2} A_2 = -j2\gamma_2 |A_1(z,t)|^2 A_2(z,t), \tag{4.207}$$

where $\gamma_2 = n_c^2 n_2 \omega_2 / n_{eff}^2(\omega_2) c A_{eff}(\omega_2)$ and the factor 2 on the right-hand side appears due to the fact that XPM effect is twice as strong as SPM. We have also assumed in Equation 4.207 that the probe signal experiences negligible SPM and has the same linear propagation loss as the pump wave. The solution for $A_2(z, t)$ is

$$A_2(z,t) = A_2(0,t)e^{-\alpha_0 z/2}e^{-j2\gamma_2|A_1(0,t)|^2 z_{eff}}, \tag{4.208}$$

where z_{eff} is the effective propagation distance given by Equation 4.57.

At the input coupling junction of the ADMR, the probe signal must satisfy the boundary condition

$$A_2(0,t) = \tau_1 A_2(L,t)e^{j\phi_{L,2}} - j\kappa_1 s_{in,2}(t), \tag{4.209}$$

where κ_1 and τ_1 are the input coupling and transmission coefficients, respectively, $L = 2\pi R$, and $\phi_{L,2} = -\beta_0(\omega_2)L$ is the linear round-trip phase of the probe beam. Substituting the solution for $A_2(L, t)$ in Equation 4.208 into the above boundary condition and solving for $A_2(0, t)$, we obtain

$$A_2(0,t) = \frac{-j\kappa_1 s_{in,2}(t)}{1 - \tau_1\tau_2 a_{rt}e^{j(\phi_{L,2}+\phi_{NL,2})}}, \tag{4.210}$$

where $a_{rt} = e^{-\alpha_0 L/2}$, $\phi_{NL,2} = -2\gamma_2|A_1(0, t)|^2 L_{eff}$, and L_{eff} is the effective microring circumference. From the solutions for $A_1(0, t)$ and $A_2(0, t)$, we can obtain the envelopes of the transmitted pump and probe pulses at the drop port and through port of the ADMR by applying Equations 4.204 and 4.205 to each of the pump and probe waves.

Figure 4.13 shows a simulation example of pump-and-probe switching in a Si ADMR with instantaneous Kerr nonlinearity. The linear parameters and Kerr coefficient of the microring are the same as those in the example of Figure 4.11. Free carrier effects are neglected. The input pump beam is a Gaussian pulse with peak power $P_0 = 1$ W, pulse width $T_0 = 50$ ps, and center frequency tuned to a microring resonance near the 1.55 µm wavelength. The probe beam is a CW signal with 1 µW average power whose frequency is tuned to a different but nearby resonance of the microring. Both the pump and probe beams are assumed to be TE polarized. The pump pulse inside the microring and those appearing at the drop port and through port of the ADMR are the same as shown in Figure 4.11b. Figure 4.13a shows the nonlinear phase change of the probe signal induced by the pump pulse inside the microring resonator. This phase change causes the probe signal to become detuned from the microring resonance. As a result, the probe power is discharged from the microring during the duration of the pump pulse, as seen in Figure 4.13b, and only returns to its resonant state after the pump pulse has passed. At the through port of the ADMR, the probe signal appears as a pulse, while at the drop port, it has the same inverted pulse shape as the probe signal inside the microring.

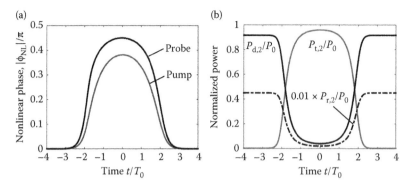

Figure 4.13 Pump-and-probe switching in a Si ADMR with instantaneous Kerr nonlinearity and negligible FC effects. The input pump beam is a Gaussian pulse with peak power $P_0 = 1$ W and pulse width $T_0 = 50$ ps. The probe beam is a CW signal with 1 μW average power. (a) Nonlinear phase changes experienced by the pump and probe signals in the microring and (b) probe signal powers at the drop port ($P_{d,2} = |s_{d,2}|^2$) and through port ($P_{t,2} = |s_{t,2}|^2$), and inside the microring ($P_{r,2} = |A_2(0,t)|^2$).

The presence of free carriers in the microring resonator can have a significant impact on the pump and probe pulses in the resonator. To simplify the analysis, we consider only the effects of TPA on the pump beam and neglect TPA of the probe beam since the peak pump power is much larger than that of the probe. Accounting for both Kerr and FC-induced nonlinearities, we first solve Equations 4.68 and 4.80 for the pump pulse envelope $A_1(z, t)$ and the average FC density $\bar{N}_{fc}(z,t)$ in the microring. The equation for the probe beam is

$$\frac{\partial A_2}{\partial z} + \frac{\alpha_0}{2} A_2 = -j2\gamma_{K,1}\,|A_1(z,t)|^2\,A_2(z,t) - j2\gamma_{fc,1}\bar{N}_{fc}A_2(z,t), \quad (4.211)$$

where $\gamma_{K,1}$ and $\gamma_{fc,1}$ are the nonlinear Kerr and FC coefficients defined in Equations 4.69 and 4.70, respectively, for the pump beam.

As a numerical example, we show in Figures 4.14a and b the pump and probe pulses, respectively, at the drop port, through port, and inside a Si ADMR. The parameters of the device are the same as those in the example of Figure 4.12. The input pump pulse has peak power $P_0 = 2.2$ mW, pulse width $T_0 = 50$ ps, and center frequency tuned to a microring resonance. The probe signal is a CW signal with 1 μW average power tuned to a nearby resonance. Due to the strong FC-induced nonlinearity in the microring, the probe signal can be

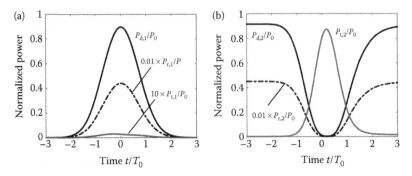

Figure 4.14 Pump-and-probe switching in a Si ADMR: plots of the powers at the drop port and through port, and in the microring for (a) the pump beam and (b) the probe beam. The input pump pulse has peak power $P_0 = 2.2$ mW and pulse width $T_0 = 50$ ps. The probe beam is a continuous wave with 1 μW average power.

switched with a much lower pump pulse power compared to the example in Figure 4.13, where only Kerr nonlinearity is present.

Experimental demonstration of pump-and-probe switching in a microring resonator was first reported in Van et al. (2002b) using free carrier effects in a vertically coupled GaAs/AlGaAs ADMR. On and off switching times of less than 100 ps were obtained, which were limited by the FC lifetime. It was also suggested that the switching speed of the device could be improved by applying a DC bias to sweep out the carriers. Other demonstrations of switching based on FC were also reported in a laterally coupled GaAs/AlGaAs microracetrack (Ibrahim et al. 2003a), as well as in a silicon microring resonator (Almeida et al. 2004). XPM was also employed to demonstrate several photonic logic gate operations with microrings, including AND/NAND gates based on InP and GaAs APMRs (Ibrahim et al. 2003b), and a NOR gate based on two cascaded GaAs/AlGaAs microring resonators (Ibrahim et al. 2004).

4.4 FWM in a Nonlinear Microring Resonator

Frequency generation based on FWM in a microring resonator was first theoretically studied in Van and Little (1999) and experimentally demonstrated in a GaAs/AlGaAs microring resonator in Absil et al. (2000). Due to the field enhancement near a microring resonance, much higher frequency conversion efficiency can be achieved in the resonator than in a straight waveguide with the same length

as the microring circumference. Applications of frequency generation in microring resonators have since been extended to frequency
comb generation (Del'Haye et al. 2008), hyperparametric oscillation
(Razzari et al. 2010), and the creation of entangled photon pairs by
spontaneous FWM (Clemmen et al. 2009).

In this section, we develop a formalism based on power coupling
in space to analyze degenerate FWM in an APMR with instantaneous Kerr nonlinearity. We consider an APMR with radius R, field
coupling coefficient κ and linear propagation loss α_0 in the microring waveguide. A pump wave with power $P_{in,p}$ and frequency ω_p
and a signal wave with power $P_{in,s}$ and frequency ω_s are applied
to the input waveguide, as depicted in Figure 4.15. For simplicity
we assume that both the pump and signal waves have the same
linear polarization. The pump and signal frequencies ω_p and ω_s
are tuned to two different resonances of the microring, with the
separation between them denoted by $\Delta\omega = \omega_s - \omega_p$. FWM generates
an idler wave at frequency $\omega_i = 2\omega_p - \omega_s = \omega_p - \Delta\omega$. If the separation
$\Delta\omega$ between the pump and signal frequencies is small (typically
equal to one FSR of the microring), the dispersion in the microring
resonances is small so that the idler frequency will also coincide
to a microring resonant frequency. We will examine the effects of
resonance detuning and phase mismatch caused by dispersion in
the microring in more detail at the end of the section.

Under the assumption of no pump depletion, we can solve the
nonlinear wave equation for the pump beam and the coupled wave
equations for the signal and idler waves separately. Defining z as
the propagation distance along the circumference of the microring
as shown in Figure 4.15, we can write the equation for the pump
wave $A_p(z)$ in the microring resonator as

$$\frac{dA_p}{dz} + \frac{\alpha_0}{2} A_p = -j\gamma |A_p|^2 A_p, \tag{4.212}$$

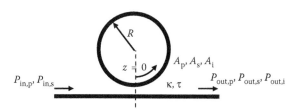

Figure 4.15   Degenerate FWM in a nonlinear APMR.

where $\gamma = \gamma(\omega_p)$ is the nonlinear coefficient in Equation 4.51. The solution of the above equation is given by Equation 4.142,

$$A_p(z) = \sqrt{P_p}\, e^{-\alpha_0 z/2} e^{-j\gamma P_p z_{eff}}, \tag{4.213}$$

where $P_p = |A_p(0)|^2$ is the pump power in the microring at $z = 0$ and $z_{eff} = (1 - e^{-\alpha_0 z})/\alpha_0$ is the effective propagation distance. At the coupling junction between the microring and the straight waveguide, the pump wave satisfies the boundary condition

$$A_p(0) = \tau A_p(L) e^{j\phi_{L,p}} - j\kappa \sqrt{P_{in,p}}, \tag{4.214}$$

where $\tau = \sqrt{1 - \kappa^2}$, $L = 2\pi R$, and $\phi_{L,p} = -n_r(\omega_p)\omega_p L/c$ is the linear round-trip phase. By substituting the solution in Equation 4.213 into the above boundary condition, we obtain an equation similar to Equation 4.145 whose solution gives the pump power P_p in the microring.

For the signal and idler waves, the coupled wave equations are given by Equations 4.129 and 4.130,

$$\frac{dA_s}{dz} + \frac{\alpha_0}{2} A_s = -j\gamma\left(2|A_p|^2 A_s + A_p^2 A_i^* e^{j\Delta\beta z}\right), \tag{4.215}$$

$$\frac{dA_i^*}{dz} + \frac{\alpha_0}{2} A_i^* = j\gamma\left(2|A_p|^2 A_i^* + (A_p^*)^2 A_s e^{-j\Delta\beta z}\right), \tag{4.216}$$

where $\Delta\beta = \beta_s + \beta_i - 2\beta_p$ is the wave vector mismatch. For simplicity we have neglected dispersion in both the linear loss and the nonlinear coefficient so that $\alpha_0(\omega_s) \approx \alpha_0(\omega_i) \approx \alpha_0$ and $\gamma(\omega_s) \approx \gamma(\omega_i) \approx \gamma$. Writing the solutions for the signal and idler waves as

$$A_s(z) = B_s(z) e^{-\alpha_0 z/2} e^{-j2\gamma P_p z_{eff}}, \tag{4.217}$$

$$A_i^*(z) = B_i^*(z) e^{-\alpha_0 z/2} e^{j2\gamma P_p z_{eff}}, \tag{4.218}$$

and making use of the pump wave solution in Equation 4.213, we can simplify Equations 4.215 and 4.216 to

$$\frac{dB_s}{dz} = -j\gamma P_p e^{-\alpha_0 z} B_i^* e^{j\psi(z)}, \tag{4.219}$$

$$\frac{dB_i^*}{dz} = j\gamma P_p e^{-\alpha_0 z} B_s e^{-j\psi(z)}, \tag{4.220}$$

where $\psi(z) = \Delta\beta z + 2\gamma P_p z_{eff}$. In general, the coupled wave equations in Equations 4.219 and 4.220 must be solved numerically. However, if we neglect the effect of attenuation of the pump beam due to linear loss in the microring, we can set $e^{-\alpha_0 z} \approx 1$ and $z_{eff} \approx z$, in which case B_s and B_i will have solutions of the form similar to Equations 4.132 and 4.133. We can thus write the solutions for A_s and A_i^* as

$$A_s(z) = [a_s \cosh(gz) + b_s \sinh(gz)]e^{-\alpha_0 z/2} e^{-j\Omega(z)}, \tag{4.221}$$

$$A_i^*(z) = [a_i^* \cosh(gz) + b_i^* \sinh(gz)]e^{-\alpha_0 z/2} e^{j\Omega(z)}, \tag{4.222}$$

where

$$g = \sqrt{(\gamma P_p)^2 - (K/2)^2}, \tag{4.223}$$

$$\Omega(z) = 2\gamma P_p z_{eff} - Kz/2, \tag{4.224}$$

$$K = \Delta\beta + 2\gamma P_p. \tag{4.225}$$

At the coupling junction, the signal and idler waves satisfy the boundary conditions

$$A_s(0) = \tau A_s(L)e^{j\phi_{L,s}} - j\kappa\sqrt{P_{in,s}}, \tag{4.226}$$

$$A_i^*(0) = \tau A_i^*(L)e^{-j\phi_{L,i}}, \tag{4.227}$$

where $\phi_{L,k} = -n_r(\omega_k)\omega_k L/c$ is the linear round-trip phase at frequency $\omega_k = \{\omega_s, \omega_i\}$. In addition, by evaluating the coupled wave equations (4.215) and (4.216) at $z = 0$, we also have the boundary conditions

$$\frac{dA_s}{dz}\bigg|_{z=0} = -\frac{\alpha_0}{2}A_s(0) - j\gamma P_p[2A_s(0) + A_i^*(0)] = gb_s - \frac{\alpha_0}{2}a_s$$
$$-j(\gamma P_p - \Delta\beta/2)a_s, \tag{4.228}$$

$$\frac{dA_i^*}{dz}\bigg|_{z=0} = -\frac{\alpha_0}{2}A_i^*(0) + j\gamma P_p[2A_i^*(0) + A_s(0)] = gb_i^* - \frac{\alpha_0}{2}a_i^*$$
$$+j(\gamma P_p - \Delta\beta/2)a_i^*. \tag{4.229}$$

Applying the boundary conditions in Equations 4.226 through 4.229 to the solutions for A_s and A_i^* in Equations 4.221 and 4.222, we solve for the coefficients a_s, b_s, a_i, and b_i to get

$$b_i^* = \frac{1 - \tau a_{rt} \cosh(gL)e^{-j\varphi_i}}{\tau a_{rt} \sinh(gL)e^{-j\varphi_i}} a_i^* = f_1 a_i^*, \tag{4.230}$$

$$a_i^* = \frac{j\gamma P_p}{gf_1 - jK/2} a_s = f_2 a_s, \tag{4.231}$$

$$b_s = -j(\gamma P_p f_2 + K/2)a_s/g = f_3 a_s, \tag{4.232}$$

$$a_s = \frac{-j\kappa\sqrt{P_{in,s}}}{1 - \tau a_{rt}[\cosh(gL) + f_3 \sinh(gL)]e^{j\varphi_s}} = f_4\sqrt{P_{in,s}}, \tag{4.233}$$

where $\varphi_s = \phi_{L,s} - \Omega(L)$, $\varphi_i = \phi_{L,i} - \Omega(L)$, and $a_{rt} = e^{-\alpha_0 L/2}$ is the round-trip amplitude attenuation due to linear loss in the microring.

Similar to the case of FWM in a straight waveguide, we find from Equation 4.223 that in order to have gain in the microring waveguide, we require $(\gamma P_p)^2 - (K/2)^2 > 0$, or $-4\gamma P_p < \Delta\beta < 0$. The maximum gain is $g_{max} = \gamma P_p$, which occurs when $K = 0$ or $\Delta\beta = -2\gamma P_p$. Since the pump power inside the microring is enhanced by approximately a factor of $FE^2 \sim 1/\kappa^2$ with respect to the input pump power, the range of wave vector mismatch for which gain is achieved becomes $-4\gamma P_{in,p}/\kappa^2 < \Delta\beta < 0$, which is broadened by a factor of FE^2 compared to FWM in a straight waveguide. However, since the circumference of a microring resonator is typically short, the phase mismatch per round-trip, $\Delta\beta L$, is small so that wave vector mismatch is less important for FWM in a microring resonator than in a straight waveguide. In fact, for values of $\Delta\beta$ outside the range $[-4\gamma P_p, 0]$ the idler wave still experiences gain per round-trip even though the waveguide gain g is imaginary.[*]

---

[*] For $\Delta\beta > 0$ or $\Delta\beta < -4\gamma P_p$, the gain is $g = jg'$ where $g' = \sqrt{(K/2)^2 - (\gamma P_p)^2}$. The solutions for the signal and idler waves are

$$A_s(z) = [a_s \cos(g'z) + b_s \sin(g'z)]e^{-\alpha_0 z/2}e^{-j\Omega(z)},$$

$$A_i^*(z) = [a_i^* \cos(g'z) + b_i^* \sin(g'z)]e^{-\alpha_0 z/2}e^{j\Omega(z)}.$$

In this case both the signal and idler waves can still exhibit gain for small propagation distance z.

The frequency conversion efficiency of the APMR can be defined as the ratio of the idler wave power at the output port to the signal wave power at the input port, $\eta_r = P_{out,i}/P_{in,s}$. The idler wave power at the output port is

$$P_{out,i} = \kappa^2 |A_i(L)|^2 = (\kappa a_{rt})^2 |a_i|^2 |\cosh(gL) + f_1 \sinh(gL)|^2 = (\kappa/\tau)^2 |a_i|^2.$$

(4.234)

The input signal power is related to the signal wave in the microring by $P_{in,s} = |a_s/f_4|^2$. Thus the conversion efficiency of the APMR is

$$\eta_r = (\kappa/\tau)^2 |f_2|^2 |f_4|^2.$$

(4.235)

In the above expression, $|f_2|^2$ is the ratio of the idler power to signal power in the microring (i.e., the conversion efficiency in the microring), and $|f_4|^2$ gives the power enhancement of the signal wave due to resonance. Under the condition of maximum gain in the microring waveguide ($K = 0$ and $g_{max} = \gamma P_p$), and assuming that the signal and idler waves are tuned to coincide with the microring resonances so that $\varphi_s = 2m_s\pi$ and $\varphi_i = 2m_i\pi$, we can evaluate the conversion efficiency of the APMR to get

$$\eta_r = \frac{\kappa^4 a_{rt}^2 \sinh^2(g_{max}L)}{\left(1 - \tau a_{rt}e^{g_{max}L}\right)^2 \left(1 - \tau a_{rt}e^{-g_{max}L}\right)^2}.$$

(4.236)

The above expression shows that when the gain g_{max} is such that $e^{g_{max}L} = 1/\tau a_{rt}$, that is, the round-trip gain completely compensates for the round-trip loss, the conversion efficiency is infinite. However, we have neglected the effect of pump depletion in our analysis, which becomes important at high conversion efficiencies. In practice, pump depletion and resonance detuning of the pump, signal, and idler waves due to SPM and XPM limit the maximum achievable conversion efficiency in the microring.

Since the product $g_{max}L$ is typically small, we can make the approximation $e^{g_{max}L} \approx e^{-g_{max}L} \approx 1$. Further assuming that the microring resonator has a high Q factor ($\tau \approx 1$ and $a_{rt} \approx 1$), we can estimate the conversion efficiency as follows:

$$\eta_r \approx \frac{\kappa^4 \sinh^2(g_{max}L)}{(1-\tau)^4} = \frac{\kappa^4(1+\tau)^4 \sinh^2(g_{max}L)}{(1-\tau^2)^4} \approx \left(\frac{16}{\kappa^4}\right)\eta_{wg},$$

(4.237)

where $\eta_{wg} = \sinh^2(g_{max}L)$ is the maximum conversion efficiency in a straight waveguide with the same length as the microring circumference. Since the field enhancement in the microring is $FE \sim 1/\kappa$, the above result shows that the conversion efficiency in an APMR is enhanced by roughly the fourth power of the field enhancement factor.

As a numerical example, we show in Figure 4.16 the FWM conversion efficiency in a Si APMR around the 1.55 µm wavelength and its dependence on various parameters. The microring is assumed to have radius $R = 20$ µm, coupling coefficient $\kappa = 0.4$, effective index $n_r = 2.5$, and linear loss coefficient $\alpha_0 = 2.5$ dB/cm. The Kerr coefficient of silicon is $n_2 = 4.5 \times 10^{-14}$ cm^2/W and the effective mode area of the waveguide is $A_{eff} = 0.15$ µm^2. Figure 4.16a plots the conversion efficiency as a function of the wave vector mismatch $\Delta\beta$ for several

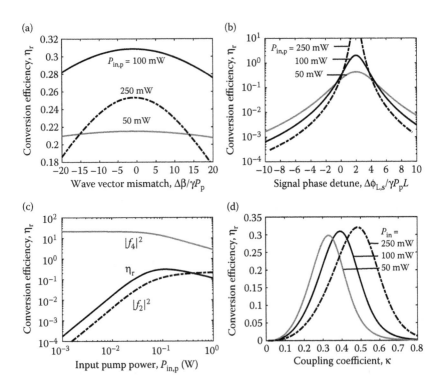

Figure 4.16 Frequency conversion efficiency (η_r) in a Si APMR: (a) η_r versus wave vector mismatch $\Delta\beta$ for $\Delta\phi_{L,p} = \Delta\phi_{L,s} = \Delta\phi_{L,i} = 0$, (b) η_r versus signal phase detune $\Delta\phi_{L,s}$ for $\Delta\beta = 0$, (c) η_r versus input pump power $P_{in,p}$ for $\Delta\beta = 0$ and $\Delta\phi_{L,p} = \Delta\phi_{L,s} = \Delta\phi_{L,i} = 0$, (d) η_r versus coupling coefficient κ for $\Delta\beta = 0$ and $\Delta\phi_{L,p} = \Delta\phi_{L,s} = \Delta\phi_{L,i} = 0$.

input pump powers. The linear phase detunes of the pump, signal, and idler waves are assumed to be zero ($\Delta\phi_{L,p} = \Delta\phi_{L,s} = \Delta\phi_{L,i} = 0$). We observe that the conversion efficiency varies only slightly with the wave vector mismatch because of the short round-trip length (circumference) of the microring. Also, the maximum conversion efficiency does not occur at $\Delta\beta = -2\gamma P_p$ as for FWM in a straight waveguide because XPM causes a slight detuning of the signal and idler waves from the microring resonances. This resonance detuning also causes a drop in the cosnversion efficiency as the input pump power is increased from 100 mW to 250 mW.

Figure 4.16b plots the conversion efficiency versus the linear phase detune of the signal wave ($\Delta\phi_{L,s}$) from the resonance for several input pump powers under the condition of zero wave vector mismatch ($\Delta\beta = 0$). We assume that the idler wave also has the same linear phase detune as the signal wave ($\Delta\phi_{L,i} = \Delta\phi_{L,s}$), whereas the pump wave has zero phase detune ($\Delta\phi_{L,p} = 0$). The plot shows that the conversion efficiency depends strongly on the linear phase detune of the signal, reaching a peak value at $\Delta\phi_{L,s} = 2\gamma P_p L$. At this linear phase detune, both the conversion efficiency inside the microring ($|f_2|^2$) and the power enhancement of the signal wave ($|f_4|^2$) are maximized, resulting in maximum idler wave power at the output of the APMR. Note that although the plot shows that the conversion efficiency exceeds 100% at high pump powers, in reality the conversion efficiency is limited by pump depletion.

Figure 4.16c shows the dependence of the conversion efficiency on the input pump power under the condition of zero wave vector mismatch ($\Delta\beta = 0$) and zero phase detunes ($\Delta\phi_{L,p} = \Delta\phi_{L,s} = \Delta\phi_{L,i} = 0$). Also shown are the variations of the conversion efficiency inside the microring ($|f_2|^2$) and the power enhancement of the signal wave ($|f_4|^2$) with the input pump power. We observe that for low pump powers, the conversion efficiency increases with the pump power due to increase in gain in the microring, reaching a maximum for $P_{in,p} \sim 0.1$ W. Above this pump power, the signal and idler waves become increasingly detuned from the microring resonances, as evident from the decrease in $|f_4|^2$, causing the conversion efficiency to drop as the pump power is further increased.

Figure 4.16d shows the dependence of the conversion efficiency on the coupling coefficient of the APMR. We observe that the conversion efficiency initially increases as κ is increased, reaching a peak value before decreasing again for larger κ values. This variation generally follows the variations of the enhancements of the pump, signal, and idler powers inside the microring with respect to the coupling coefficient.

We now examine the effect of waveguide dispersion on the resonance detuning of the idler wave and the phase-matching condition inside the microring. We assume that the pump beam and the signal wave are tuned exactly to two different resonant frequencies, Ω_p and Ω_s, respectively, of the microring:

$$\omega_p = \Omega_p = \frac{m_p}{n_r(\Omega_p)} \frac{c}{R}, \tag{4.238}$$

$$\omega_s = \Omega_s = \frac{m_s}{n_r(\Omega_s)} \frac{c}{R} = \omega_p + \Delta\omega, \tag{4.239}$$

where $n_r(\omega)$ is the effective index of the microring waveguide, m_p and m_s are the resonance mode numbers, and $\Delta\omega = \omega_s - \omega_p$ is the frequency separation between the pump and the signal wave. The idler wave is generated at frequency $\omega_i = 2\omega_p - \omega_s$, which is detuned from the resonance mode Ω_i of the microring by

$$\delta\omega_i = \omega_i - \Omega_i = (2\omega_p - \omega_s) - \frac{m_i}{n_r(\Omega_i)} \frac{c}{R}, \tag{4.240}$$

where $m_i = 2m_p - m_s$. Writing the above equation as

$$\delta\omega_i = 2\left[\omega_p - \frac{m_p c}{n_r(\Omega_p)R} \frac{n_r(\Omega_p)}{n_r(\Omega_i)}\right] - \left[\omega_s - \frac{m_s c}{n_r(\Omega_s)R} \frac{n_r(\Omega_s)}{n_r(\Omega_i)}\right], \tag{4.241}$$

we see that if there is no waveguide dispersion, $n_r(\Omega_p) = n_r(\Omega_s) = n_r(\Omega_i) = n_r$, $\delta\omega_i$ is zero, and the idler wave coincides exactly with the resonance mode m_i of the microring ($\omega_i = \Omega_i$). However, due to dispersion, the idler wave will be slightly detuned from the microring resonance Ω_i. To determine the amount of detuning, we approximate the effective index of the microring waveguide in terms of a first-order Taylor series expansion around the pump beam frequency,

$$n_r(\omega) = n_r(\Omega_p) + (\omega - \Omega_p)D_1, \tag{4.242}$$

where $D_1 = dn_r/d\omega$ evaluated at $\omega = \Omega_p$. The derivative D_1 can be related to the group index of the microring waveguide through the formula $n_g = n_r(\Omega_p) + \Omega_p D_1$. Using the above Taylor series approximation, we can express the index ratios $n_r(\Omega_p)/n_r(\Omega_i)$ and $n_r(\Omega_s)/n_r(\Omega_i)$ in Equation 4.241 as

$$\frac{n_r(\Omega_p)}{n_r(\Omega_i)} = \frac{n_r(\Omega_p)}{n_r(\Omega_p) + (\Omega_i - \Omega_p)D_1} \approx 1 - (\Omega_i - \Omega_p)D_1/n_r, \qquad (4.243)$$

$$\frac{n_r(\Omega_s)}{n_r(\Omega_i)} = \frac{n_r(\Omega_p) + (\Omega_s - \Omega_p)D_1}{n_r(\Omega_p) + (\Omega_i - \Omega_p)D_1} \approx 1 + (\Omega_s - \Omega_i)D_1/n_r, \qquad (4.244)$$

where $n_r = n_r(\Omega_p)$. With the above approximations, Equation 4.241 can be simplified to

$$\delta\omega_i \approx \left[2\Omega_p(\Omega_i - \Omega_p) + \Omega_s(\Omega_s - \Omega_i) \right] D_1/n_r. \qquad (4.245)$$

Substituting $\Omega_s = \Omega_p + \Delta\omega$ and $\Omega_i = \Omega_p - \Delta\omega - \delta\omega_i$ into Equation 4.245 and solving for $\delta\omega_i$, we obtain the following expression for the resonance detune (or frequency walk-off) of the idler wave:

$$\delta\omega_i \approx \frac{2D_1\Delta\omega^2}{n_g - D_1\Delta\omega}. \qquad (4.246)$$

The above equation shows that the amount of resonance detuning experienced by the idler wave increases as the square of the frequency separation $\Delta\omega$ between the pump and signal waves. As the idler wave is detuned from the microring resonance, the conversion efficiency decreases since the wave is no longer enhanced by resonance. Resonance detuning due to waveguide dispersion ultimately limits the maximum achievable frequency conversion bandwidth, or gain bandwidth, of FWM in a microring resonator.

Next we examine how the phase-matching condition is affected by dispersion in the microring waveguide. The wave vector mismatch $\Delta\beta$ is given by

$$\Delta\beta = \beta(\omega_s) + \beta(\omega_i) - 2\beta(\omega_p). \qquad (4.247)$$

We can express the propagation constant in terms of a Taylor series around the pump frequency ω_p as (Marhic et al. 1996)

$$\beta(\omega) = \beta(\omega_p) + \sum_{n=1}^{\infty} \frac{\beta_n}{n!}(\omega - \omega_p)^n, \qquad (4.248)$$

where $\beta_n = (\partial^n \beta / \partial \omega^n)_{\omega = \omega_p}$. Upon substituting Equation 4.248 into 4.247, we find that all the odd-order terms cancel out and only the even-order terms contribute to the wave vector mismatch:

$$\Delta\beta = \Delta\omega^2 \beta_2 + \frac{1}{12}\Delta\omega^4 \beta_4 + \cdots. \tag{4.249}$$

Thus to the lowest order, the wave vector mismatch also increases as the square of the frequency separation $\Delta\omega$ between the pump and signal waves. However, since the round-trip length of a microring resonator is typically short, the total round-trip phase mismatch $\Delta\beta L$ caused by dispersion is small so that phase mismatch plays a less critical role than resonance detuning in limiting the FWM conversion bandwidth of an APMR.

Finally we note that, in a semiconductor microring resonator with significant TPA, FCD and FCA can also have deleterious effects on the FWM conversion efficiency and bandwidth (Lin et al. 2007). It was reported in Ong et al. (2013) that by actively removing the generated carriers using a reverse-biased pin junction, a twofold improvement in the FWM conversion efficiency in a silicon microring resonator could be achieved.

4.5   Summary

This chapter reviews important nonlinear optics applications based on an intensity-dependent nonlinear refractive index in a microring resonator. These applications include bistable devices, self-oscillations, all-optical switching, and frequency conversion. In many of these applications, the large field enhancement near a microring resonance helps amplify the nonlinear effects, leading to a reduction in the required input power and hence better device efficiency compared to similar processes in a straight waveguide. For each application, a formalism based on power coupling in space is also developed to provide a semi-analytical analysis of the nonlinear optical process involved. While only intensity-dependent nonlinear refraction and absorption are considered in this chapter, the formalisms developed can also be applied to analyze microring devices with other types of nonlinearity, such as second-order nonlinearity, Raman scattering, and Brillouin scattering.

References

Absil, P. P., Hryniewicz, J. V., Little, B. E., Cho, P. S., Wilson, R. A., Joneckis, L. G., Ho, P.-T. 2000. Wavelength conversion in GaAs micro-ring resonators. *Opt. Lett.* 25: 554–556.

Agrawal, G. P. 2013. *Nonlinear Fiber Optics*, 5th Ed. Oxford: Academic Press.

Almeida, V. R., Barrios, C. A., Panepucci, R. R., Lipson, M. 2004. All-optical control of light on a silicon chip. *Nature* 431(28): 1081–1084.

Armaroli, A., Malaguti, S., Bellanca, G., Trillo, S., de Rossi, A., Combrié, S. 2011. Oscillatory dynamics in nanocavities with noninstantaneous Kerr response. *Phy. Rev. A* 84(5): 053816.

Boyd, R. W. 2008. *Nonlinear Optics*, 3rd Ed. New York: Academic Press.

Chen, S., Zhang, L., Fei, Y., Cao, T. 2012. Bistability and self-pulsation phenomena in silicon microring resonators based on nonlinear optical effects. *Opt. Express* 20(7): 7454–7468.

Clemmen, S., Phan H. K., Bogaerts, W., Baets, R. G., Emplit, Ph., Massar, S. 2009. Continuous wave photon pair generation in silicon-on-insulator waveguides and ring resonators. *Opt. Express* 17(19): 16558–16570.

Del'Haye, P., Arcizet, O., Schliesser, A., Holzwarth, R., Kippenberg, T. J. 2008. Full stabilization of a microresonator-based optical frequency comb. *Phys. Rev. Lett.* 101: 053903.

Dimitropoulos, D., Jhaveri, R., Claps, R., Woo, J. C. S., Jalali, B. 2005. Lifetime of photogenerated carriers in silicon-on-insulator rib waveguides. *Appl. Phys. Lett.* 86: 071115.

Dinu, M., Quochi, F., Garcia, H. 2003. Third-order nonlinearities in silicon at telecom wavelengths. *Appl. Phys. Lett.* 82(18): 2954–2956.

Eggleton, B. J., Moss, D. J., Radic, S. 2008. *Optical Fiber Telecommunications V: Components and Sub-systems* (eds. I. P. Kaminow, T. Li, A. E. Willner), Chapter 20, pp. 759–828. San Diego: Academic Press.

Ibrahim, T. A., Amarnath, K., Kuo, L. C., Grover, R., Van, V., Ho, P.-T. 2004. Photonic logic NOR gate based on two symmetric microring resonators. *Opt. Lett.* 29(23): 2779–2781.

Ibrahim, T. A., Cao, W., Kim, Y., Li, J., Goldhar, J., Ho P.-T., Lee, C. H. 2003a. All-optical switching in a laterally coupled microring resonator by carrier injection. *IEEE Photonics Technol. Lett.* 15(1): 36–38.

Ibrahim, T. A., Grover, R., Kuo, L.-C., Kanakaraju, S., Calhoun, L. C., Ho, P.-T. 2003b. All-optical AND/NAND logic gates using semiconductor microresonators. *IEEE Photonics Technol. Lett.* 15(10): 1422–1424.

Ikeda, K., Akimoto, O. 1982. Instability leading to periodic and chaotic self-pulsations in a bistable optical cavity. *Phys. Rev. Lett.* 48(9): 617–620.

Ikeda K., Daido, H. 1980. Optical turbulence: Chaotic behavior of transmitted light from a ring cavity. *Phys. Rev. Lett.* 45(9): 709–712.

Ikeda, K., Saperstein, R. E., Alic, N., Fainman, Y. 2008. Thermal and Kerr nonlinear properties of plasma-deposited silicon nitride/silicon dioxide. *Opt. Express* 16(17): 12987–12994.

Islam, M. N., Soccolich, C. E., Slusher, R. E., Levi, A. F. J., Hobson, W. S., Young, M. G. 1992. Nonlinear spectroscopy near half-gap in bulk and quantum well GaAs/AlGaAs waveguides. *J. App. Phys.* 71(4): 1927–1935.

Johnson, T. J., Borselli, M., Painter, O. 2006. Self-induced optical modulation of the transmission through a high-Q silicon microdisk resonator. *Opt. Express* 14: 817–831.

Lin, Q., Painter, O. J., Agrawal, G. P. 2007. Nonlinear optical phenomena in silicon waveguides: Modeling and applications. *Opt. Express* 15(25): 16604–16644.

Malaguti, S., Bellanca, G., de Rossi, A., Combrie, S., Trillo, S. 2011. Self-pulsing driven by two-photon absorption in semiconductor cavities. *Phys. Rev. A* 83: 051802(R).

Marhic, M. E., Kagi, N., Chiang, T.-K., Kazovsky, L. G. 1996. Broadband fiber optical parametric amplifiers. *Opt. Lett.* 21(8): 573–575.

Marcuse, D. 1991. *Theory of Dielectric Optical Waveguides*, 2nd Ed. Boston: Academic Press.

Ong, J. R., Kumar, R., Aguinaldo, R., Mookherjea, S. 2013. Efficient CW four-wave mixing in silicon-on-insulator micro-rings with active carrier removal. *IEEE Photonics Technol. Lett.* 25(17): 1699–1702.

Pernice, W. H. P., Li, M., Tang, H. X. 2010. Time-domain measurement of optical transport in silicon micro-ring resonators. *Opt. Express* 18: 18438–18452.

Priem, G., Dumon, P., Bogaerts, W., Van Thourhout, D., Morthier, G., Baets, R. 2005. Optical bistability and pulsating behaviour in silicon-on-insulator ring resonator structures. *Opt. Express* 13: 9623–9628.

Razzari, L., Duchesne, D., Ferrera, M., Morandotti, R., Chu, S., Little, B. E., Moss, D. J. 2010. CMOS-compatible integrated optical hyper-parametric oscillator. *Nat. Photonics* 4: 41–45.

Silberberg, Y., Bar-Joseph, I. 1984. Optical instabilities in a nonlinear Kerr medium. *J. Opt. Soc. Am. B* 1(4): 662–670.

Soref, R., Bennett, B. 1987. Electrooptical effects in silicon. *IEEE J. Quantum Electron.* 23: 123–129.

Turner-Foster, A. C., Foster, M. A., Levy, J. S., Poitras, C. B., Salem, R., Gaeta, A. L., Lipson, M. 2010. Ultrashort free-carrier lifetime in low-loss silicon nanowaveguides. *Opt. Express* 18(4): 3582–3591.

Van, V., Ibrahim, T. A., Absil, P. P., Johnson, F. G., Grover, R., Ho, P.-T. 2002a. Optical signal processing using nonlinear semiconductor microring resonators. *IEEE J. Sel. Top. Quantum Electron.* 8(3): 705–713.

Van, V., Ibrahim, T. A., Ritter, K., Absil, P. P., Johnson, F. G., Grover, R., Goldhar, J., Ho, P.-T. 2002b. All-optical nonlinear switching in GaAs-AlGaAs microring resonators. *IEEE Photonics Technol. Lett.* 14(1): 74–77.

Van, V., Little, B. E. 1999. Design and analysis of nonlinear microring resonators for third-order harmonic generation. In *Integrated Photonics Research Conference*, Santa Barbara, CA, paper RTuB3.

Zhang, L., Fei, Y., Cao, T., Cao, Y., Xu, Q., Chen, S. 2013. Multibistability and self-pulsation in nonlinear high-Q silicon microring resonators considering thermo-optical effect. *Phys. Rev. A* 87: 053805.

Zhang, L., Fei, Y., Cao, Y., Lei, X., Chen, S. 2014. Experimental observations of thermo-optical bistability and self-pulsation in silicon microring resonators. *J. Opt. Soc. Am. B* 31(2): 201–206.

CHAPTER 5

Active Photonic Applications of Microring Resonators

In Chapter 4, we showed that the resonant frequencies of a microring resonator can be modified by the electric field of the lightwave circulating inside the resonator through the nonlinear optical processes of self-phase modulation and cross phase modulation. In this chapter, we consider microring devices in which the resonant frequency can be tuned by applying an electrical signal such as a voltage or a current. The ability to electrically tune the microring resonances allows a variety of important active photonic device applications to be realized such as tunable filters, switches, and modulators.

This chapter reviews the principles and technologies for achieving tuning, switching, and modulation of a microring resonator by modifying its resonant frequencies. Section 5.1 gives an overview of the common physical mechanisms by which the resonant frequencies of a microring resonator can be tuned, namely, the thermo-optic effect, the electro-optic effect, and free carrier dispersion. The dynamic response of a microring modulator under the small-signal condition will be analyzed in Section 5.2. Section 5.3 extends the analysis to large signals by taking into account the nonlinear transfer function of the microring modulator. Important device performance characteristics such as modulation efficiency, bandwidth, and nonlinearity will also be discussed.

5.1 Mechanisms for Tuning Microring Resonators

In general, to tune the resonant wavelengths of a microring resonator, we need to change the effective index n_{eff} of the microring waveguide. Suppose the effective index of a section of length L of the microring is changed by Δn_{eff}, for example, by the application of a voltage V_r^* we can determine the wavelength shift $\Delta\lambda$ experienced by resonance mode λ_m as follows. We begin by writing the resonance condition as

$$(2\pi R - L)n_{\text{eff}}(\lambda) + Ln_{\text{eff}}(\lambda, V) = m\lambda_m, \tag{5.1}$$

where we have separated the microring circumference into a section of length L whose effective index depends on both the wavelength and the applied voltage, and a section of length $(2\pi R - L)$ whose effective index is not affected by the voltage. Taking the differential change of Equation 5.1 with respect to both the wavelength and the applied voltage, we get

$$(2\pi R - L)\frac{dn_{\text{eff}}}{d\lambda}\Delta\lambda + L\left(\frac{\partial n_{\text{eff}}}{\partial \lambda}\Delta\lambda + \frac{\partial n_{\text{eff}}}{\partial V}V\right) = m\Delta\lambda, \tag{5.2}$$

which can be simplified to give

$$2\pi R\left(\frac{dn_{\text{eff}}}{d\lambda} - \frac{n_{\text{eff}}}{\lambda_m}\right)\Delta\lambda + L\frac{\partial n_{\text{eff}}}{\partial V}V = 0, \tag{5.3}$$

where we have made use of the relation $m = 2\pi Rn_{\text{eff}}/\lambda_m$. Using the expression for the group index, $n_g = n_{\text{eff}} - \lambda_m dn_{\text{eff}}/d\lambda$, we obtain from Equation 5.3

$$\Delta\lambda = \frac{\lambda_m}{n_g}\frac{L}{2\pi R}\Delta n_{\text{eff}}(V), \tag{5.4}$$

where $\Delta n_{\text{eff}}(V) = (\partial n_{\text{eff}}/\partial V)V$ is the index change due to the applied voltage alone.

We can vary the effective index of the microring waveguide by changing the refractive index of the core material, the cladding, or both. If Δn_c and Δn_{cl} are the index changes in the core and cladding material, respectively, the resulting change in the effective index can be evaluated from the expression

$$\Delta n_{\text{eff}} = \frac{2n_c\Delta n_c\Gamma_c}{n_{\text{eff}}} + \frac{2n_{cl}\Delta n_{cl}\Gamma_{cl}}{n_{\text{eff}}}, \tag{5.5}$$

---

* The analysis also holds if the voltage is replaced by any other external means of changing the effective index, such as temperature, pressure, or electric field in the case of nonlinear optical waveguides.

where n_c and n_{cl} are the original indices of the core and cladding, respectively. The factors Γ_c and Γ_{cl} represent the fraction of the mode power residing in the core and cladding, respectively, and are given by

$$\Gamma_i = \frac{\int_{S_i} |\phi(x,y)|^2 \, dx \, dy}{\int |\phi(x,y)|^2 \, dx \, dy}, \tag{5.6}$$

where $i = \{c, cl\}$, $\phi(x, y)$ is the field distribution of the waveguide mode, and S_i is the cross-sectional area of the core or cladding.

The changes in the refractive indices of the core and cladding materials can be effected by several mechanisms, the most common being the thermo-optic effect, the electro-optic effect, and free carrier dispersion in the waveguide. We briefly discuss each of these tuning mechanisms below.

5.1.1 The thermo-optic effect

The thermo-optic effect describes the change in the refractive index of a material with temperature, which can be expressed by the formula

$$n(T) = n(T_0) + \alpha(T - T_0), \tag{5.7}$$

where $\alpha = dn/dT$ is the thermo-optic coefficient. For example, the thermo-optic coefficients of Si and SiO$_2$ at room temperature are 1.86×10^{-4} K^{-1} (Cocorullo et al. 1999) and 9×10^{-6} K^{-1} (Leviton and Frey 2006), respectively. In general, the thermo-optic coefficients themselves are also functions of the temperature.

Since both the core and cladding materials of a waveguide exhibit the thermo-optic effect, the variation in the effective index of the waveguide with temperature will be the result of the thermo-optic changes in the indices of both the core and cladding materials, as given by Equation 5.5. Some materials, such as SU8 and PMMA (polymethylmethacrylate) polymers, have negative thermo-optic coefficients. These materials can be used as the cladding to compensate for the positive thermo-optic change in the refractive index of the core material in order to minimize the temperature sensitivity of the device, or even to achieve athermal device operation (Raghunathan et al. 2010).

A simple way to achieve thermo-optic tuning of a microring resonator (or any integrated optics device) is by varying the temperature of the substrate using a Peltier cooler or heater. This method is useful for controlling the temperature of the entire chip. For more localized thermo-optic control of a microring, a micro-heater is typically fabricated directly above the structure, as shown in Figure 5.1. Micro-heaters are typically made of highly resistive materials, such as metals (e.g., Ni/Cr [Sherwood-Droz et al. 2008], Ti [Dong et al. 2010a]). By passing a current through the heater element, resistive dissipation causes local heating of the waveguide section directly underneath the heater. The temperature change experienced by the waveguide can be determined by solving the heat diffusion equation

$$\rho C \frac{\partial T}{\partial t} = \nabla \cdot (K \nabla T) + q, \tag{5.8}$$

over the waveguide cross section in the x–y plane, as shown in Figure 5.1b. In the above equation, $T(x, y)$ is the temperature distribution, ρ, C, and K are the mass density, specific heat capacity, and thermal conductivity, respectively, of the materials comprising the waveguide. The heat source q can be equated to the power density dissipated from the resistor, $q = J^2/\sigma$, where J is the current density and σ is the electric conductivity of the heater. Under steady-state condition, Equation 5.8 reduces to Poisson's equation, $\nabla \cdot (K \nabla T) = -q$.

From the temperature distribution, we can calculate the thermo-optic change in the refractive index $n(x, y)$ over the waveguide cross section, which then allows us to determine the new effective index

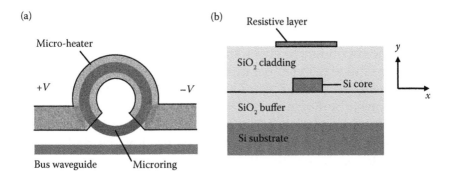

Figure 5.1   Schematic of a micro-heater on a silicon microring resonator: (a) top view and (b) cross-sectional view.

of the waveguide. Typically, the effective index has a roughly linear dependence on the power and can be expressed by the relation $\Delta n_{eff} = \eta P/L$, where η is the thermo-optic tuning rate, and P is the total power dissipated over the length L of the micro-heater. Using Equation 5.4, we find that the shift in the resonant wavelengths of a microring resonator of radius R is given by

$$\Delta\lambda = \frac{\eta P}{2\pi R} \frac{\lambda_m}{n_g}. \tag{5.9}$$

The above expression shows that the resonant wavelength shift is directly proportional to the dissipated power and inversely proportional to the microring radius. Thus, for the same amount of power dissipation, we obtain a larger wavelength shift in a microring with a smaller radius. However, the power required to tune a resonant wavelength across the full FSR of a microring is relatively independent of its radius.* Note also that the resistor length L does not appear in Equation 5.9 since it is the total power dissipated, and not the power dissipated per unit length, that determines the resonant wavelength shift.

In designing micro-heaters, consideration should be paid to the optimal placement of the heater above the waveguide so as to maximize the temperature rise with minimum power consumption, while avoiding excess optical loss due to light absorption in the conductor. Another design issue is thermal crosstalk, which is important when several components are placed in close proximity to each other. An example is the thermo-optic tuning of individual microrings in a coupled microring device. In such a structure, heating one microring may also cause residual heating of adjacent resonators, resulting in unwanted resonant shifts in the latter. Thus, the heaters must be carefully designed and positioned in order to minimize thermal crosstalk.

The time response of the thermo-optic effect is limited by thermal diffusion, which is a relatively slow process, typically on the scale of micro or milliseconds. For this reason, thermo-optic tuning is more suitable for applications in which high-speed tuning or modulation of the microring resonance is not required.

----

* Since the FSR of a microring resonator is given by $\Delta\lambda_{FSR} = \lambda^2/2\pi n_g R$, the power required to tune a resonant wavelength over the full FSR is $P = (\Delta\lambda_{FSR} 2\pi R/\eta) \times (n_g/\lambda) = \lambda/\eta$, which is independent of the microring radius.

Some application examples include thermo-optic switches in a WDM switch matrix, stabilization of the resonance wavelength of a microring resonator, and alignment of microring resonances in high-order microring filters.

Owing to the large thermo-optic coefficient of silicon, it is possible to achieve efficient thermal tuning of silicon microring resonators. For example, for a silicon waveguide with core dimensions 250×450 nm^2 lying on a 1-μm-thick buffer oxide, simulation shows that using a NiCr heater with 2 μm width and 100 nm thickness located 1 μm above the waveguide, we can achieve a tuning rate of about $\eta = 4.3 \times 10^{-2}$ μm/mW in terms of the effective index change of the TE mode. Figure 5.2a shows the temperature distribution over the waveguide cross section when the power dissipation is 0.1 mW per μm of heater length. For a 10-μm-radius silicon microring resonator, assuming that the micro-heater covers the entire microring circumference, the tuning rate is around 20 mW/FSR shift in the microring resonant wavelength. Figure 5.2b shows the tuning curves (wavelength shift vs. power) of the silicon microring for different heater positions above the silicon waveguide. We observe that the tuning curves are relatively linear, with the tuning rate increasing as the heater is placed closer to the microring waveguide. In general, the tuning rate depends on various device parameters such as the dimensions of the waveguide and the micro-heater, the buffer oxide thickness, the location of the heater above the waveguide, and the polarization of the input light.

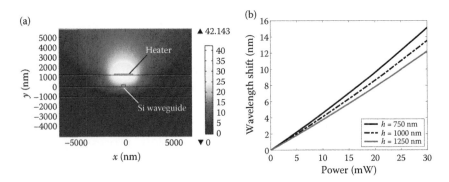

Figure 5.2 (a) Temperature distribution (degree celsius) over a silicon waveguide cross section for a 0.1 mW of power dissipation per μm heater length and (b) resonant wavelength shift of a 10-μm-radius silicon microring as a function of the dissipated power for heater position (h) of 750 nm, 1000 nm, and 1250 nm above the silicon waveguide.

Experimentally, reported tuning rates for silicon microring resonators are in the range of 10–50 mW/FSR (Sherwood-Droz et al. 2008, Dong et al. 2010a,b).

5.1.2 The electro-optic effect

The electro-optic effect refers to the change in the refractive index of a material caused by an applied electric field. In contrast to the nonlinear optic effects discussed in Chapter 4, the electric field here is assumed to be static or slowly varying compared to the relaxation time of the material. The formalism for describing the electro-optic effect is based on the optical indicatrix (or index ellipsoid), which maps the index experienced by a wave propagating in a given direction in a crystal. In the absence of an applied field, the equation for the index ellipsoid is

$$\frac{x^2}{n_x^2} + \frac{y^2}{n_y^2} + \frac{z^2}{n_z^2} = 1, \tag{5.10}$$

where n_x, n_y, and n_z are the indices seen by a wave polarized along the principal optical axes (x, y, z) of the crystal. The refractive index seen by a wave with the electric field polarized in a given direction is found from the intersection of the ellipsoid with the line drawn from the origin and in the direction parallel to the polarization vector.

An applied electric field has the effect of rotating the index ellipsoid, so that the equation for the new ellipse assumes the more general form (following the convention in Yariv (1991))

$$\left(\frac{1}{n^2}\right)_1 x^2 + \left(\frac{1}{n^2}\right)_2 y^2 + \left(\frac{1}{n^2}\right)_3 z^2 + 2\left(\frac{1}{n^2}\right)_4 yz + 2\left(\frac{1}{n^2}\right)_5 zx$$
$$+ 2\left(\frac{1}{n^2}\right)_6 xy = 1. \tag{5.11}$$

The coefficients $(1/n^2)_i$ are optical constants describing the new index ellipsoid. They can be related to the coefficients $(1/n^2)_i^{(0)}$ of the ellipsoid in the absence of the applied field by

$$\left(\frac{1}{n^2}\right)_i = \left(\frac{1}{n^2}\right)_i^{(0)} + \Delta\left(\frac{1}{n^2}\right)_i, \tag{5.12}$$

where $\Delta(1/n^2)_i$ represents the change in the coefficients due to the applied field. In the absence of the applied field, Equation 5.11 reduces to Equation 5.10, so we must have

$$\left(\frac{1}{n^2}\right)_1^{(0)} = \frac{1}{n_x^2}; \quad \left(\frac{1}{n^2}\right)_2^{(0)} = \frac{1}{n_y^2}; \quad \left(\frac{1}{n^2}\right)_3^{(0)} = \frac{1}{n_z^2},$$

$$\left(\frac{1}{n^2}\right)_4^{(0)} = \left(\frac{1}{n^2}\right)_5^{(0)} = \left(\frac{1}{n^2}\right)_6^{(0)} = 0.$$

For the linear electro-optic effect (also known as the Pockels effect), the changes in the optical constants are directly proportional to the applied field and are given by

$$\Delta\left(\frac{1}{n^2}\right)_i = \sum_{j=1}^{3} r_{i,j} E_j, \tag{5.13}$$

where $r_{i,j}$ are the linear electro-optic coefficients and the subscript $j = \{1, 2, 3\}$ corresponds to field directions $\{x, y, z\}$. Equation 5.13 can be written in tensor form as

$$\begin{bmatrix} \Delta(1/n^2)_1 \\ \Delta(1/n^2)_2 \\ \Delta(1/n^2)_3 \\ \Delta(1/n^2)_4 \\ \Delta(1/n^2)_5 \\ \Delta(1/n^2)_6 \end{bmatrix} = \begin{bmatrix} r_{11} & r_{12} & r_{13} \\ r_{21} & r_{22} & r_{23} \\ r_{31} & r_{32} & r_{33} \\ r_{41} & r_{42} & r_{43} \\ r_{51} & r_{52} & r_{53} \\ r_{61} & r_{62} & r_{63} \end{bmatrix} \begin{bmatrix} E_1 \\ E_2 \\ E_3 \end{bmatrix} = \mathbf{r}:\mathbf{E}, \tag{5.14}$$

where $\mathbf{r}$ is the linear electro-optic tensor. The linear electro-optic effect occurs only in crystals with no inversion symmetry, such as GaAs, AlGaAs, GaP, and LiNbO$_3$. On the other hand, centrosymmetric crystals such as Si exhibit the quadratic electro-optic effect. Even in crystals lacking inversion symmetry, rotational symmetry considerations further lead to a reduction in the number of unique nonzero elements in the electro-optic tensor. For example, the electro-optic tensor of GaAs has only three nonzero elements, $r_{41} = r_{52} = r_{63} = 1.43$ pm/V, at the 1.15 μm wavelength (Yariv 1991). The equation for the index ellipsoid of a GaAs crystal in the presence of an applied electric field $\mathbf{E} = E_x\hat{\mathbf{x}} + E_y\hat{\mathbf{y}} + E_z\hat{\mathbf{z}}$ is

$$\frac{1}{n_0^2}x^2 + \frac{1}{n_0^2}y^2 + \frac{1}{n_0^2}z^2 + 2r_{41}E_x yz + 2r_{41}E_y zx + 2r_{41}E_z xy = 1, \quad (5.15)$$

where $n_0 = 3.4$ is the refractive index of GaAs when there is no applied field.

The refractive index of a crystal in the presence of an applied field is found from the intersection of the new ellipsoid with the line drawn from the origin and in the direction parallel to the wave's polarization. For example, suppose a GaAs crystal is subject to an electric field in the x direction, $(\mathbf{E} = E_x \hat{x})$, the equation for the new index ellipsoid is

$$\left(\frac{1}{n_0^2}\right)x^2 + \left(\frac{1}{n_0^2}\right)y^2 + \left(\frac{1}{n_0^2}\right)z^2 + 2r_{41}E_x yz = 1. \quad (5.16)$$

Making the coordinate transformations

$$X = x,$$
$$Y = y\cos(45°) + z\sin(45°) = (y+z)/\sqrt{2},$$
$$Z = -y\sin(45°) + z\cos(45°) = (z-y)/\sqrt{2},$$

we can write Equation 5.16 as

$$\frac{X^2}{n_0^2} + \left(\frac{1}{n_0^2} - r_{41}E_x\right)Y^2 + \left(\frac{1}{n_0^2} + r_{41}E_x\right)Z^2 = 1. \quad (5.17)$$

By equating

$$\frac{1}{n_Y^2} = \frac{1}{n_0^2} - r_{41}E_x,$$

$$\frac{1}{n_Z^2} = \frac{1}{n_0^2} + r_{41}E_x,$$

we find that the refractive indices seen by a wave polarized in the Y and Z directions, respectively, are given by

$$n_Y \approx n_0 + \frac{1}{2}n_0^3 r_{41}E_x, \quad (5.18)$$

$$n_Z \approx n_0 - \frac{1}{2}n_0^3 r_{41}E_x. \quad (5.19)$$

Note that the index seen by a wave polarized in the X direction is $n_X = n_0$, indicating that the wave is unaffected by the applied field. This example demonstrates that the electro-optic effect is highly anisotropic and can even vanish for certain directions of polarization.

In electro-optic modulator applications, an important parameter commonly used to quantify the strength of the nonlinear effect is V_π, which is the voltage required to achieve a π phase shift in the lightwave. For example, for a GaAs waveguide of length L placed between two electrodes separated by distance d, the voltage required to induce a π phase shift in the lightwave is $V_\pi = \lambda d / n_0^3 r_{41} L$.

Since the electro-optic effect has its origin in the response of bound electrons to the applied electric field, it is characterized by a very fast response time, typically in the order of 10^{-14} s (Agullo-Lopez et al. 1994). For this reason, the electro-optic effect is commonly employed in high-speed modulator applications. The switching speed of an electro-optic modulator is limited not by the material response time, but rather by the parasitic RC time constant of the electrical connections, as well as the velocity mismatch between the electrical and optical signals.

Materials exhibiting the electro-optic effect are not restricted to only crystalline solids. Certain classes of polymers are also known to possess very large electro-optic coefficients. For example, a phenyltetraene-bridged chromophore (labeled CLD) in an amorphous polycarbonate (APC) host material has an electro-optic coefficient of 55 pm/V at the 1.55 μm wavelength (Oh et al. 2000), which is almost 40 times larger than that in GaAs. This polymer material has also been used to fabricate efficient microring modulators operating in the telecommunication wavelength range (Rabiei et al. 2002).

5.1.3   Free carrier dispersion

The effective index of a semiconductor waveguide can also be modulated by varying the free carrier concentration in the waveguide. The change in the refractive index and optical absorption of a semiconductor material due to the presence of N_e electron concentration and N_h hole concentration are given by Equations 4.16 and 4.17

$$\Delta n = \sigma_r^{(e)} N_e + \sigma_r^{(h)} N_h, \tag{5.20}$$

$$\Delta \alpha = \sigma_a^{(e)} N_e + \sigma_a^{(h)} N_h, \tag{5.21}$$

where $\sigma_r^{(e,h)}$ and $\sigma_a^{(e,h)}$ are the refraction volume and absorption cross section, respectively, due to electron (e) and hole (h) concentrations. The refraction volume and absorption cross section may be evaluated by treating the electrons and holes as plasmas embedded in a background dielectric of index n_0. The relative permittivity of the semiconductor in the presence of excess electrons or holes can be written as

$$\varepsilon_r + \Delta\varepsilon_r = (n_0 + \Delta n - j\Delta\alpha)^2 \approx n_0^2 + 2n_0(\Delta n - j\Delta\alpha), \tag{5.22}$$

where $\varepsilon_r = n_0^2$. From the Drude model for a plasma, the change in the relative permittivity $\Delta\varepsilon_r$ due to electron or hole concentration $N_{(e,h)}$ is given by

$$\Delta\varepsilon_r \approx -\frac{N_{(e,h)}q^2}{\varepsilon_0 m^*_{(e,h)}} \frac{1}{\omega(\omega - j\gamma)}, \tag{5.23}$$

where $m^*_{(e,h)}$ is the effective mass of an electron or hole in the semiconductor, γ is the free carrier relaxation rate, and q is the elementary charge. Substituting Equation 5.23 into Equation 5.22 and using the relations in Equations 5.20 and 5.21, we get

$$\sigma_r^{(e,h)}(\omega) \approx -\frac{q^2}{2n_0\varepsilon_0 m^*_{(e,h)}}\left(\frac{1}{\omega^2 + \gamma^2}\right), \tag{5.24}$$

$$\sigma_a^{(e,h)}(\omega) \approx \frac{q^2}{2n_0\varepsilon_0 m^*_{(e,h)}}\left(\frac{\gamma/\omega}{\omega^2 + \gamma^2}\right). \tag{5.25}$$

For silicon around the 1.55 μm wavelength, it is more accurate to use the empirical values for $\sigma_r^{(e,h)}$ and $\sigma_a^{(e,h)}$ given in Equations 4.20 and 4.21.

Free carriers are typically induced in a semiconductor waveguide by the application of a voltage across a semiconductor junction. A variety of semiconductor junctions have been employed to achieve modulation of the free carrier density, including Schottky junction (Campbell et al. 1975), pn junction (Dong et al. 2009), pin junction (Xu et al. 2005, Li et al. 2007), metal–oxide–semiconductor capacitor (MOS CAP) (Liu et al. 2004), and silicon–insulator–silicon capacitor (SISCAP) (Milivojevic et al. 2013). Figure 5.3 shows the schematics of silicon waveguide structures embedded with a

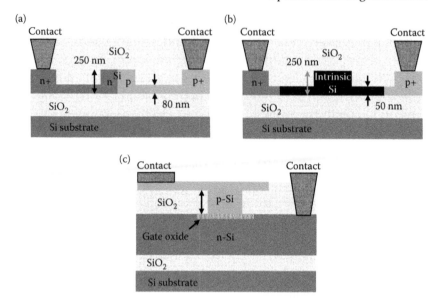

Figure 5.3 Schematic of (a) a horizontal pn junction silicon waveguide (After Dong, P., et al., 2009, *Opt. Express*, 17, (25): 22484–22490), (b) a pin junction silicon waveguide (After Xu, Q., et al., 2005, *Nature*, 435, (19): 325–327), and (c) an MOS capacitor silicon rib waveguide (After Liu, A., et al., 2004, *Nature*, 427, (12): 615–618).

pn junction, a pin junction, and an MOS capacitor. Typically, pin junctions are operated in the forward bias mode (carrier injection), while pn junctions can be operated in either the forward bias or reverse bias mode (depletion mode). Likewise, MOS capacitors can be operated in either the carrier depletion mode or carrier accumulation mode. In general, the forward bias mode provides larger changes in the free carrier densities and hence larger changes in the refractive index than the reverse bias mode. However, it is also inherently slower since a larger amount of charge must be moved in and out of the junction. For this reason, high-speed modulators typically employ depletion-mode pn junctions or depletion-mode MOS capacitors. The switching or modulation speed of these junctions is determined by the recombination time and transit time of the free carriers, as well as the RC time constant of the electrical contacts.

The relationship between the induced free carrier concentrations and applied voltage in a semiconductor junction waveguide is determined by solving Poisson's equation of electrostatics and the charge transport equation in the junction (Sze 1981). Commercial

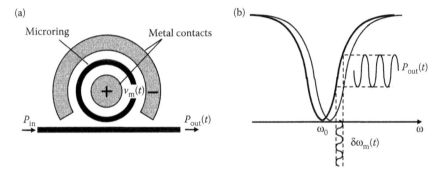

Figure 5.4 (a) Schematic of an all-pass microring modulator with constant input light power P_{in} and modulating voltage $v_m(t)$ applied across the microring waveguide and (b) plot of the microring resonance spectrum showing modulation in the transmitted power $P_{out}(t)$ due to variation in the resonant frequency $\delta\omega_m(t)$.

software such as Atlas, Crosslight, Lumerical, and Comsol, are also available which can be used to obtain accurate simulations of the charge–voltage characteristics and the time-domain response of charge transport in practical device structures. Electrical circuit models of the junctions can also be constructed to assist in the design and optimization of switches and modulators based on these semiconductor junction waveguides.

5.2 Dynamic Response of a Microring Modulator

In this section, we develop a set of equations based on the energy coupling formalism for analyzing the dynamic response of an APMR due to a change in the resonant frequency (Pile and Taylor 2014). These equations are general in that they are not specific to the underlying mechanism used to effect the resonant frequency shift, and can be used to model the dynamic response of a microring in switching and modulation applications. Since these applications typically involve small shifts in the resonant frequency compared to the FSR of the resonator, the energy coupling formalism will provide sufficient accuracy for our analysis.[*]

The principle of operation of an APMR switch or intensity modulator is illustrated in Figure 5.4. An input monochromatic

---

[*] An analysis of the dynamic response of a microring modulator based on the power coupling formalism can be found in Sacher and Poon (2008).

lightwave is tuned to a frequency on the shoulder of the resonance spectrum of the microring resonator. Suppose a voltage signal $v_m(t)$ is applied across the microring waveguide which induces a change in its effective index. This causes a shift $\delta\omega_m$ in the resonance spectrum of the microring resulting in a corresponding change in the transmitted light power, $P_{out}(t)$, of the microring resonator.

In Section 5.2.1, we will formulate a set of equations based on energy coupling which govern the time-domain response of an APMR modulator. We will then solve these equations in the small-signal (SS) approximation in Section 5.2.2, where the modulator response is linearized around a bias point. Important characteristics of the modulator such as modulation efficiency, bandwidth, and step response will be discussed. Extension of the analysis to large-signal (LS) condition, which takes into account the nonlinear characteristic of the microring transfer function, will be treated in Section 5.3.

5.2.1   Dynamic energy-coupling model of a microring modulator

We consider an APMR with resonant frequency ω_0, energy coupling coefficient μ, and intrinsic decay rate γ_0 due to loss. The intrinsic decay rate can be computed from the microring waveguide loss α as $\gamma_0 = \alpha v_g/2$, where v_g is the group velocity. For an optical signal $\tilde{s}_{in}(t)$ applied to the input port of the APMR, the equation for the energy-normalized wave amplitude $\tilde{a}(t)$ in the microring is

$$\frac{d\tilde{a}}{dt} = (j\omega_0 - \gamma_r)\tilde{a}(t) - j\mu\tilde{s}_{in}(t), \tag{5.26}$$

where $\gamma_r = \gamma_0 + \gamma_e$ is the total decay rate and $\gamma_e = \mu^2/2$ is the extrinsic decay rate due to coupling to the bus waveguide. In the following analysis, we assume that the input light is a monochromatic wave with frequency ω and power P_{in} so that $\tilde{s}_{in}(t) = \sqrt{P_{in}}\,e^{j\omega t}$. Supposing that the resonant frequency of the microring experiences a time-dependent change given by $\delta\omega_m(t)$, the instantaneous microring resonant frequency is $\omega_0 + \delta\omega_m$. Writing $\tilde{a}(t) = a(t)e^{j\omega t}$, we obtain from Equation 5.26

$$\frac{da}{dt} = -(j\Delta\omega + \gamma_r)a + j\delta\omega_m a - j\mu\sqrt{P_{in}}, \tag{5.27}$$

where $\Delta\omega = \omega - \omega_0$ is the frequency detune of the input light signal from the microring resonance.

The change in the microring resonant frequency is related to the change in the effective index δn_r of the microring waveguide by $\delta\omega_m(t) = -\omega_0\delta n_r(t)/n_g$, where n_g is the group index. We will assume that the effective index can be changed through the application of a voltage v_m to the modulator, and that the relationship between the index change and the applied voltage is linear, $\delta n_r = \eta v_m$, where η is the refractive index change per volt. We will further assume that the response time of the effective index to the applied voltage is characterized by a time constant τ_m, which we call the index modulation time constant. This time constant depends on the specific mechanism used to induce the index change, the physical structure and dimensions of the device, and the parasitic capacitances and resistances associated with the electrical contacts. With the above assumptions, we can express the time dependence of the resonant frequency modulation on the applied voltage as

$$\frac{d\delta\omega_m}{dt} = -\frac{\delta\omega_m}{\tau_m} - \left(\frac{\eta\omega_0}{n_g\tau_m}\right)v_m(t). \tag{5.28}$$

If the response of the effective index to the applied voltage is instantaneous, then Equation 5.28 simply gives $\delta\omega_m(t) = -(\eta\omega_0/n_g)v_m(t)$. The solution of Equations 5.27 and 5.28 gives the dynamic response of the wave amplitude $a(t)$ in the microring. The transmitted light signal at the output of the APMR can then be determined from the relation $s_{out}(t) = \sqrt{P_{in}} - j\mu a(t)$.

The above equations can also be applied to a microring modulator or switch in the add-drop configuration. For an ADMR with input and output energy coupling coefficients μ_1 and μ_2, respectively, the total decay rate in Equation 5.27 is given by $\gamma_r = \gamma_0 + \mu_1^2/2 + \mu_2^2/2$ and the input term is modified to read $-j\mu_1\sqrt{P_{in}}$. The transmitted light signals at the drop port and through port of the microring are given by the relations $s_d(t) = -j\mu_2 a(t)$ and $s_t(t) = \sqrt{P_{in}} - j\mu_1 a(t)$, respectively.

In the next section, we will solve Equations 5.27 and 5.28 for an APMR modulator under the SS approximation to obtain the relationship between the output modulated light power and the applied voltage.

5.2.2 Small-signal analysis

In the absence of an applied modulating signal ($\delta\omega_m = 0$), the wave amplitude in the microring is constant, $a(t) = a_0$, with the solution given by

$$a_0 = \frac{-j\mu\sqrt{P_{in}}}{j\Delta\omega + \gamma_r} = \frac{-j\mu\sqrt{P_{in}}}{\Gamma}, \tag{5.29}$$

where $\Gamma = j\Delta\omega + \gamma_r$. Suppose a small modulating signal $v_m(t)$ is applied to the microring causing a small resonant frequency shift $\delta\omega_m(t)$. As a result, the wave amplitude in the microring deviates from the steady-state value by a small amount $\delta a_m(t)$ so that we can write $a(t) = a_0 + \delta a_m(t)$. Making this substitution in Equation 5.27 and keeping only terms linear in $\delta\omega_m(t)$ and $\delta a_m(t)$, we get

$$\frac{d\delta a_m}{dt} = -\Gamma\delta a_m + ja_0\delta\omega_m. \tag{5.30}$$

Taking the Fourier transform of the above equation and solving for δa_m, we obtain

$$\delta a_m(j\Omega) = \frac{ja_0}{j\Omega + \Gamma}\delta\omega_m(j\Omega), \tag{5.31}$$

where Ω represents the electrical (or RF) modulation frequency, and $\delta a_m(j\Omega)$ and $\delta\omega_m(j\Omega)$ denote the Fourier transform of the wave amplitude in the microring and the resonant frequency modulation, respectively. The resonant frequency spectrum $\delta\omega_m(j\Omega)$ can be obtained from the Fourier transform of Equation 5.28 as

$$\delta\omega_m(j\Omega) = -\left(\frac{\eta\omega_0}{n_g\tau_m}\right)\frac{V_m(j\Omega)}{j\Omega + 1/\tau_m}, \tag{5.32}$$

where $V_m(j\Omega)$ denotes the Fourier transform of the modulating voltage signal. Substituting Equation 5.32 into Equation 5.31, we get

$$\delta a_m(j\Omega) = -\left(\frac{\eta\omega_0}{n_g\tau_m}\right)\frac{ja_0 V_m(j\Omega)}{(j\Omega + \Gamma)(j\Omega + 1/\tau_m)}. \tag{5.33}$$

Since the transmitted light signal at the output of the APMR is given by $\tilde{s}_{out}(t) = \tilde{s}_{in} - j\mu\tilde{a}(t)$, we can determine the output modulated light power from

$$P_{out}(t) = \left|\tilde{s}_{out}(t)\right|^2 = \left|\sqrt{P_{in}} - j\mu a(t)\right|^2. \tag{5.34}$$

Substituting $a(t) = a_0 + \delta a_m(t)$ into the above equation, we get

$$P_{out}(t) = \left| \sqrt{P_{in}} - j\mu(a_0 + \delta a_m) \right|^2 = \left| s_{out} - j\mu \delta a_m \right|^2, \tag{5.35}$$

where s_{out} is the steady-state (or DC) transmitted light amplitude in the absence of modulation

$$s_{out} = \sqrt{P_{in}} - j\mu a_0 = (1 - \mu^2/\Gamma)\sqrt{P_{in}}. \tag{5.36}$$

We can expand Equation 5.35 to read

$$P_{out}(t) = \left| s_{out} \right|^2 + j\mu \delta a_m^* s_{out} - j\mu \delta a_m s_{out}^* + \mu^2 \left| \delta a_m \right|^2 = P_{out}^{DC} + P_{out}^{AC}(t), \tag{5.37}$$

where P_{out}^{DC} is the steady-state (or DC) transmitted light power in the absence of modulation

$$P_{out}^{DC} = \left| s_{out} \right|^2 = \left| 1 - \mu^2/\Gamma \right|^2 P_{in}, \tag{5.38}$$

and $P_{out}^{AC}(t)$ is the AC component of the transmitted light power due to modulation

$$P_{out}^{AC}(t) = j\mu(\delta a_m^* s_{out} - \delta a_m s_{out}^*) + \mu^2 \left| \delta a_m \right|^2. \tag{5.39}$$

Upon substituting the expression for s_{out} in Equation 5.36 into Equation 5.39 and neglecting the small contribution $\mu^2 |\delta a_m|^2$, we get

$$P_{out}^{AC}(t) = j\mu \sqrt{P_{in}} \left[\left(1 - \frac{\mu^2}{\Gamma} \right) \delta a_m^*(t) - \left(1 - \frac{\mu^2}{\Gamma^*} \right) \delta a_m(t) \right]. \tag{5.40}$$

Taking the Fourier transform of the above equation, we have

$$P_{out}^{AC}(j\Omega) = j\mu \sqrt{P_{in}} \left[\left(1 - \frac{\mu^2}{\Gamma} \right) \delta a_m^*(-j\Omega) - \left(1 - \frac{\mu^2}{\Gamma^*} \right) \delta a_m(j\Omega) \right]. \tag{5.41}$$

By substituting the expression for $\delta a_m(j\Omega)$ in Equation 5.33 into the above equation, we obtain the frequency response of the modulator due to an applied RF signal $V_m(j\Omega)$.

5.2.2.1 Transfer function of an APMR modulator

The transfer function of an APMR modulator can be obtained by applying an impulse voltage signal $v_m(t) = \delta(t)$ to the device and determining its frequency response. Since the Fourier transform of the impulse signal is $V_m(j\Omega) = 1$, we get from Equation 5.33

$$\delta a_m(j\Omega) = -\left(\frac{\eta\omega_0}{n_g\tau_m}\right)\frac{\mu\sqrt{P_{in}}/\Gamma}{(j\Omega+\Gamma)(j\Omega+1/\tau_m)}. \tag{5.42}$$

Substituting the above result into Equation 5.41, we obtain the transfer function of the modulator as

$$H_m(j\Omega) = P_{out}^{AC}(j\Omega) = \frac{K_0(j\Omega+2\gamma_0)}{(j\Omega+\Gamma)(j\Omega+\Gamma^*)(j\Omega+1/\tau_m)}, \tag{5.43}$$

$$K_0 = -\frac{2\Delta\omega\mu^2}{\Delta\omega^2+\gamma_r^2}\left(\frac{\eta\omega_0}{n_g\tau_m}\right)P_{in}.$$

Equation 5.43 indicates that the modulator has a third-order response characterized by three poles and one zero. Two of the poles are related to the microring resonance through the complex frequency parameter Γ, and the third pole arises from the time response of the waveguide index to the applied voltage, which is characterized by the index modulation time constant τ_m.

5.2.2.2 Modulation efficiency

The modulation efficiency of an APMR modulator can be defined as the modulation gain at DC, $G = H_m(0)$, which is the slope of the transmitted light power versus applied voltage curve at DC. Evaluating Equation 5.43 at $\Omega = 0$, we get

$$G = H_m(0) = \frac{2\gamma_0 K_0}{|\Gamma|^2(1/\tau_m)} = -\frac{4\Delta\omega\gamma_0\mu^2}{(\Delta\omega^2+\gamma_r^2)^2}\left(\frac{\eta\omega_0}{n_g}\right)P_{in}. \tag{5.44}$$

Note that the DC gain is independent of the index modulation time constant τ_m. It is convenient to express Equation 5.44 in terms of normalized parameters as follows:

$$G = -16Q_0\left(\frac{\eta P_{in}}{n_g}\right)\frac{\delta}{(1+\delta^2)^2}\frac{\gamma}{(1+\gamma)^3}, \tag{5.45}$$

where $\delta = \Delta\omega/\gamma_r$ is the normalized frequency detune, $\gamma = \gamma_e/\gamma_0$ is the ratio of the extrinsic decay rate to the intrinsic decay rate, and $Q_0 = \omega_0/2\gamma_0$ is the intrinsic Q factor of the resonator. The normalized frequency detune δ gives the ratio of the frequency detune to the half width at half max (HWHM) bandwidth of the microring, $\delta = \Delta\omega/\Delta\omega_{3dB}$ (since the HWHM bandwidth $\Delta\omega_{3dB} = \gamma_r$). The parameter γ can also be regarded as the normalized coupling because it is equal to the energy coupling normalized by the intrinsic loss, $\gamma = \mu^2/\alpha v_g$.

From Equation 5.45, we find that for a given intrinsic Q_0 factor of the microring, the maximum gain, G_{max}, occurs at $\delta = 1/\sqrt{3}$ and $\gamma = 1/2$ and is equal to

$$G_{max} = -\frac{4Q_0}{3\sqrt{3}}\left(\frac{\eta P_{in}}{n_g}\right). \tag{5.46}$$

If we normalize the gain G in Equation 5.45 by G_{max}, we obtain the expression

$$\frac{G}{G_{max}} = \frac{12\sqrt{3}\delta}{(1+\delta^2)^2}\frac{\gamma}{(1+\gamma)^3}, \tag{5.47}$$

which is independent of the microring loss. In general, for a given value of γ, we obtain the highest DC gain if the normalized frequency detune is set to be $\delta = 1/\sqrt{3}$. On the other hand, for a fixed frequency detune δ, we obtain the maximum DC gain when the normalized coupling parameter γ is equal to 1/2.

Figure 5.5a and b plot the normalized DC gain (G/G_{max}) as a function of the normalized frequency detune δ and the normalized coupling γ. In Figure 5.5a, we see that the DC gain initially increases with the frequency detune, reaching a peak value at $\delta = 1/\sqrt{3} \approx 0.58$ before decreasing again with further increase in the frequency detune. Note that the choice of δ also determines the optical insertion loss of the APMR modulator. At $\delta = 1$, which corresponds to an insertion loss of 3 dB, the DC gain is about 1.1 dB below the peak gain value.

Figure 5.5b shows a similar dependence of the DC gain on the normalized coupling parameter, with the peak gain value reached at $\gamma = 1/2$. At critical coupling $(\gamma = 1)$, the DC gain is about 0.7 dB below the peak gain value. The plot also indicates that in general, to obtain high modulation efficiency, the microring resonator should be operated in the undercoupling regime $(\gamma < 1)$.

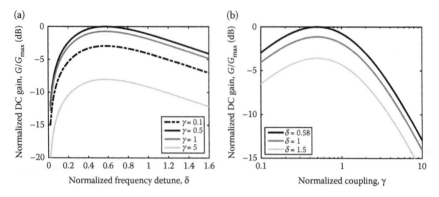

Figure 5.5 (a) Plot of the normalized DC gain (G/G_{max}) versus the normalized frequency detune δ for various values of the coupling parameter γ and (b) plot of the normalized DC gain versus the normalized coupling γ for several values of the frequency detune δ.

5.2.2.3 Electrical bandwidth

To determine the electrical (or RF) bandwidth of an APMR modulator, we consider two separate cases: the case where the modulator response is dominated by the index modulation time constant τ_m, and the case where it is dominated by the microring resonance. In the first case, the transfer function of the modulator can be simplified by making the approximation $|\Gamma| \gg \Omega$ in Equation 5.43. This results in the expression

$$H_m(j\Omega) = \frac{G/\tau_m}{j\Omega + 1/\tau_m},$$

(5.48)

where the DC gain G is given by Equation 5.44. Equation 5.48 indicates that the modulator has a first-order frequency response, with the 3 dB bandwidth of the magnitude response, $|H_m(j\Omega)|$, given by $\Omega_{3dB} = \sqrt{3}/\tau_m$.

For the case where the response of the modulator is dominated by the microring resonance, we make the approximation $1/\tau_m \gg \Omega$ in Equation 5.43 to obtain the following transfer function for the modulator:

$$H_m(j\Omega) = \frac{G|\Gamma|^2}{2\gamma_0} \frac{(j\Omega + 2\gamma_0)}{(j\Omega + \Gamma)(j\Omega + \Gamma^*)}.$$

(5.49)

The above transfer function can also be expressed in terms of the normalized frequency detune and normalized coupling as

$$H_m(j\tilde{\Omega}) = \frac{G(1+\gamma)(1+\delta^2)[j\tilde{\Omega}+2/(1+\gamma)]}{2[j\tilde{\Omega}+(1+j\delta)][j\tilde{\Omega}+(1-j\delta)]}, \tag{5.50}$$

where $\tilde{\Omega} = \Omega/\gamma_r = \Omega/\Delta\omega_{3dB}$ is the normalized RF frequency. The normalized 3 dB RF bandwidth, $\tilde{\Omega}_{3dB} = \Omega_{3dB}/\Delta\omega_{3dB}$, of $|H_m(j\tilde{\Omega})|$ can be computed from

$$\tilde{\Omega}_{3dB} = \sqrt{b + \sqrt{b^2 + 3(1+\delta^2)^2}}, \tag{5.51}$$

$$b = \frac{1}{2}(1+\gamma)^2(1+\delta^2)^2 - (1-\delta^2).$$

When the modulator is operated at maximum gain G_{max} ($\gamma = 1/2$ and $\delta = 1/\sqrt{3}$), the normalized RF bandwidth is equal to $\tilde{\Omega}_{3dB} = 2$.

The plots in Figure 5.6a and b show the frequency response of the normalized gain $|H_m(j\tilde{\Omega})/G|$ of an APMR modulator for the case where the index modulation time constant τ_m can be neglected. Figure 5.6a plots the frequency responses for a fixed coupling parameter $\gamma = 1/2$ and various values of the normalized frequency detune, while Figure 5.6b shows the responses for a fixed frequency detune $\delta = 1/\sqrt{3} \approx 0.58$ and various values of the coupling parameter. We observe that the electrical bandwidth of the modulator in general increases with the frequency detune and the coupling parameter. The variations of the electrical bandwidth with respect to δ and γ are further illustrated in Figure 5.6c and d. The increase in the electrical bandwidth with frequency detune can be explained by the fact that the charging time of the microring resonator decreases as we move farther away from the microring resonance. The microring charging time can also be decreased by increasing the external coupling, which lowers the Q factor of the resonator, with the result that the electrical bandwidth increases with larger γ.

It is also of interest to examine the trade-off between the modulation efficiency and the electrical bandwidth of an APMR modulator. Figure 5.7a plots the relationship between the normalized DC gain (G/G_{max}) and the normalized RF bandwidth ($\tilde{\Omega}_{3dB}$) at several fixed values of δ by parameterizing γ. Figure 5.7b plots the same relationship by parameterizing δ at several fixed values of γ. In both

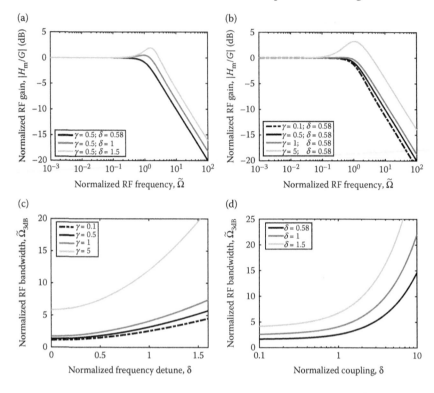

Figure 5.6 Frequency response of the normalized gain $|H_m(j\tilde{\Omega})/G|$ of an APMR modulator for (a) fixed coupling parameter $\gamma = 1/2$ and various values of δ, (b) fixed frequency detune $\delta = 1/\sqrt{3}$ and various values of γ, (c) plot of the normalized RF bandwidth versus frequency detune δ for various values of γ, and (d) plot of the normalized RF bandwidth versus coupling parameter γ for various values of δ.

plots, we see that the maximum DC gain is achieved for a normalized RF bandwidth of 2. Further increase in the RF bandwidth beyond this value leads to a corresponding decrease in the DC gain.

The bandwidth limitation of a microring modulator due to resonance (or the finite cavity lifetime) can be overcome by modulating the coupling coefficient rather than the round-trip phase of the APMR (Sacher and Poon 2008, Sacher et al. 2013). The transmitted light of the APMR is thus modulated by varying the extrinsic Q factor of the resonator. In this case, the RF gain of the modulator does not roll off at high frequencies and so, the electrical bandwidth is not limited by the linewidth of the microring resonance. The tradeoffs of this modulation scheme, however, are the smaller modulation efficiency and larger device size compared to phase-modulated

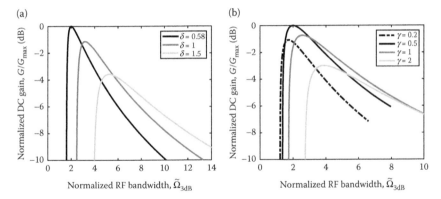

Figure 5.7 Plots of the normalized DC gain (G/G_{max}) versus the normalized RF bandwidth ($\tilde{\Omega}_{3dB}$) of an APMR modulator by (a) varying γ while keeping δ fixed and (b) by varying δ while keeping γ fixed.

APMRs (Pile and Taylor 2014). The electrical bandwidth of a coupling-modulated APMR is ultimately limited by the index modulation time constant τ_m.

5.2.2.4 Small-signal step response

The SS response of an APMR modulator to a step change in the voltage can be obtained by solving Equation 5.30 in the time domain. Suppose the applied voltage undergoes a step change from 0 to a constant value V_0 at time $t = 0$. Assuming that the effective index of the microring waveguide changes instantaneously with the applied voltage (i.e., $\tau_m = 0$), the resonant frequency of the microring also experiences an instantaneous step change equal to $\Delta\omega_m = -(\eta\omega_0/n_g)V_0$. The solution of Equation 5.30 in this case is given by

$$\delta a_m(t) = \frac{ja_0\Delta\omega_m}{\Gamma}(1 - e^{-\Gamma t}). \tag{5.52}$$

The AC transmitted power of the modulator is found using Equation 5.40*

$$P_{out}^{AC}(t) = V_0 G\left\{1 - [\cos(\Delta\omega t) - B\sin(\Delta\omega t)]e^{-\gamma_r t}\right\}, \tag{5.53}$$

---

* Note that Equation (5.40) neglects the nonlinear term $\mu^2|\delta a_m|^2$, which is proportional to $\Delta\omega_m^2$ (or V_0^2). Since this term is always positive regardless of the sign of V_0, inclusion of this term will result in unequal positive and negative swings of the transmitted power $P_{out}^{AC}(t)$. We omit this term in the SS analysis but will include it in the LS analysis in Section 5.3.1.

$$B = \frac{\Delta\omega^2 + \gamma_e^2 - \gamma_0^2}{\Delta\omega/2\gamma_0},$$

where G is the DC gain in Equation 5.44. The above result shows that the transmitted power has a second-order time response characterized by oscillations at frequency $\Delta\omega$ and decay rate γ_r. Following the conventional description of a second-order linear system, we can characterize the SS response of the modulator as overdamped if $\Delta\omega/\gamma_r = \delta < 1$, underdamped if $\Delta\omega/\gamma_r > 1$, and critically damped if $\Delta\omega/\gamma_r = 1$. Equation 5.53 also indicates that the swing in the steady-state (or DC) transmitted power, $\Delta P_{out} = P_{out}^{AC}(t \to \infty) = V_0 G$, is linearly proportional to the applied voltage V_0 (or to the resonant frequency shift $\Delta\omega_m$). The modulation index of the APMR modulator can be determined from $\eta_m = V_0 G / P_{out}^{DC}$, where P_{out}^{DC} is given by Equation 5.38.

5.3 Large-Signal Response of a Microring Modulator

Since the spectral response of an APMR near a resonance is highly nonlinear, the transfer function of the microring modulator is also highly nonlinear. For large modulating signals, this nonlinearity manifests itself through effects such as saturation and harmonic distortion. In this section, we will investigate the LS behaviors of an APMR modulator for the cases of input step voltage and sinusoidal excitations. In both cases, we will assume that the effective index of the microring responds instantaneously to the applied voltage so that we can neglect the index modulation time constant τ_m.

5.3.1 Large-signal step response

To obtain the LS response of an APMR modulator to an applied step voltage, we solve Equation 5.27 directly in the time domain. Similar to the SS analysis in the previous section, we assume that the input voltage undergoes a step change from 0 to V_0 at time $t = 0$, which causes a corresponding shift in the resonant frequency of the microring given by $\Delta\omega_m = -(\eta\omega_0/n_g)V_0$. The general solution of Equation 5.27 for the wave amplitude in the microring comprises of a forced response term and a natural response term, $a(t) = a_f + a_n(t)$. The forced response a_f is obtained when the modulator has reached a new steady state due to the resonant frequency shift $\Delta\omega_m$. Setting $da_f/dt = 0$ in Equation 5.27, we get

$$-\Gamma a_f + j\Delta\omega_m a_f - j\mu\sqrt{P_{in}} = 0,$$

which can be solved for a_f to give

$$a_f = \frac{-j\mu\sqrt{P_{in}}}{\Gamma - j\Delta\omega_m}. \tag{5.54}$$

The natural response $a_n(t)$ is the solution to the homogeneous equation

$$\frac{da_n}{dt} = -\Gamma a_n,$$

which yields

$$a_n(t) = A_0 e^{-\Gamma t}. \tag{5.55}$$

To determine the constant A_0, we note that at $t = 0$, just before the step voltage is applied, the wave amplitude in the microring is given by $a(0) = a_0 = -j\mu\sqrt{P_{in}}/\Gamma$. Using this initial condition to find A_0, we obtain the following solution for the total response $a(t)$:

$$a(t) = \frac{j\mu\sqrt{P_{in}}}{\Gamma - j\Delta\omega_m}\left(\frac{j\Delta\omega_m}{\Gamma}e^{-\Gamma t} - 1\right). \tag{5.56}$$

To determine the transmitted light power at the modulator output, we write the solution for the wave amplitude as $a(t) = a_0 + \delta a_m(t)$, where the LS modulation $\delta a_m(t)$ is given by

$$\delta a_m(t) = \Delta a_m(1 - e^{-\Gamma t}), \tag{5.57}$$

$$\Delta a_m = \frac{\Delta\omega_m \mu\sqrt{P_{in}}}{\Gamma(\Gamma - j\Delta\omega_m)}.$$

By substituting $\delta a_m(t)$ into Equation 5.39, we obtain the LS modulation $P_{out}^{AC}(t)$ of the transmitted light power, which can be expressed as

$$P_{\text{out}}^{\text{AC}}(t) = V_0 G_{\text{LS}} \left\{ 1 + \left[C_1 \cos(\Delta\omega t) + C_2 \sin(\Delta\omega t) \right] e^{-\gamma_r t} + C_3 e^{-2\gamma_r t} \right\}, \quad (5.58)$$

$$C_1 = -\frac{2\Delta\omega - (\gamma_r/\gamma_0)\Delta\omega_m}{2\Delta\omega - \Delta\omega_m},$$

$$C_2 = \frac{\Delta\omega(\Delta\omega + \Delta\omega_m) + \gamma_e^2 - \gamma_0^2}{\gamma_0(2\Delta\omega - \Delta\omega_m)},$$

$$C_3 = \frac{(\gamma_e/\gamma_0)\Delta\omega_m}{2\Delta\omega - \Delta\omega_m}.$$

In the above equation, the LS DC gain G_{LS} is given by

$$G_{\text{LS}} = -\frac{2\mu^2\gamma_0(2\Delta\omega - \Delta\omega_m)}{(\Delta\omega^2 + \gamma_r^2)[(\Delta\omega + \Delta\omega_m)^2 + \gamma_r^2]} \left(\frac{\eta\omega_0}{n_r} \right) P_{\text{in}}. \quad (5.59)$$

Note that since $\Delta\omega_m = -(\eta\omega_0/n_g)V_0$, G_{LS} depends on the applied voltage V_0 as a result of the nonlinear transmission curve of the microring resonator. In the limit of a vanishingly small resonant frequency shift ($\Delta\omega_m \to 0$), G_{LS} reduces to the SS DC gain G in Equation 5.44. The swing in the steady-state (or DC) transmitted power is given by $\Delta P_{\text{out}} = P_{\text{out}}^{\text{AC}}(t \to \infty) = V_0 G_{\text{LS}}$. We also note that the LS response contains a term which decays at a rate of $2\gamma_r$. This fast-decaying term comes from the second-order correction $\mu_2|\delta a_m|^2$, which is neglected in the SS analysis (see also footnote on page 261).

Figure 5.8 compares the transmitted power swings under the LS condition and SS condition by plotting $\Delta P_{\text{out}}/P_{\text{in}}$ versus the normalized resonant frequency shift $\Delta\omega_m/\gamma_r$ for two sets of microring parameters: the maximum gain case with $\gamma = 1/2$ and $\delta = 1/\sqrt{3}$, and the critical coupling case with $\gamma = 1$ and $\delta = 1$. As expected, the SS transmission curves are linear for both cases but with different gain slopes. The LS curves agree with the SS curves for small frequency shifts ($|\Delta\omega_m/\gamma_r| < 0.1$), but become highly nonlinear and deviate from the SS curves at larger frequency shifts. We also observe that the LS curves have asymmetric positive and negative signal swings.

The plots in Figure 5.9 show the time-domain responses ($P_{\text{out}}^{\text{AC}}/P_{\text{in}}$ vs. $\gamma_r t$) of an APMR modulator to step changes in the resonant frequency of magnitudes $\Delta\omega_m/\gamma_r = \pm 0.1$ and $\Delta\omega_m/\gamma_r = \pm 0.2$. The microring is assumed to be critically coupled ($\gamma = 1$) and the frequency detune δ is adjusted to show the cases of overdamped (plot a), critically damped (plot b), and underdamped response

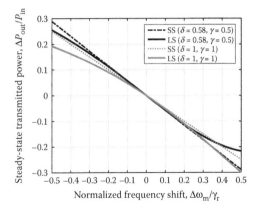

Figure 5.8   Plot of the steady-state transmitted power swing, $\Delta P_{out}/P_{in}$, versus the normalized resonant frequency shift, $\Delta \omega_m/\gamma_r$, for two sets of microring parameters: the maximum gain case with $\gamma = 1/2$ and $\delta = 1/\sqrt{3}$, and the critical coupling case with $\gamma = 1$, $\delta = 1$. Solid lines are LS solutions; dotted lines are SS solutions.

(plot c). Both LS responses (solid lines) and SS responses (dotted lines) are shown for comparison, with the deviations between the two solutions becoming more noticeable at larger step changes in the resonant frequency. As expected, the time-domain responses for the underdamped case are characterized by faster rise times and larger overshoots than the overdamped case.

5.3.2   Large-signal response under sinusoidal modulation

When the microring resonant frequency is modulated by a large sinusoidal RF signal, nonlinearity in the modulator response leads to the generation of high-order harmonic frequencies which cause distortion of the output modulated signal. We consider again Equation 5.26 with an input monochromatic lightwave at frequency ω and power P_{in}, $\tilde{s}_{in} = \sqrt{P_{in}}\,e^{j\omega t}$. Suppose the resonant frequency of the microring is modulated by a sinusoidal RF signal at frequency Ω,

$$\delta \omega_m(t) = \Delta \omega_m \cos(\Omega t) = \Delta \omega_m (e^{j\Omega t} + e^{-j\Omega t})/2, \qquad (5.60)$$

where $\Delta \omega_m$ is the amplitude of the resonant frequency shift. Owing to the nonlinear transfer function of the modulator, new harmonics are generated and mix with each other, so we expect the wave amplitude in the microring to consist of an infinite series of harmonics in general. However, for computational purposes,

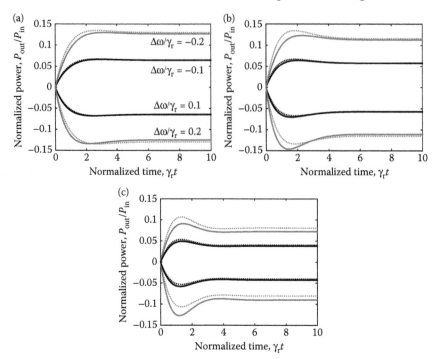

Figure 5.9   Plots of the step response of an APMR modulator showing the normalized modulated power (P_{out}^{AC}/P_{in}) versus the normalized time ($\gamma_r t$) due to a step change in the resonant frequency: (a) overdamped response with $\delta = 1/\sqrt{3}$, $\gamma = 1$, (b) critically damped response with $\delta = 1$; $\gamma = 1$, and (c) underdamped response with $\delta = 1.2$; $\gamma = 1$. The resonant frequency shifts shown in each plot are $\Delta\omega_m/\gamma_r = \pm 0.1$ and ± 0.2. Solid lines are LS solutions and dotted lines are SS solutions.

we will limit the series to the first N harmonics so that $\tilde{a}(t)$ can be expressed as

$$\tilde{a}(t) = \sum_{n=-N}^{N} a_n e^{j(\omega + n\Omega)t}. \tag{5.61}$$

Substituting the above expressions for $\delta\omega_m(t)$ and $\tilde{a}(t)$ into Equation 5.26, with ω_0 replaced by $\omega_0 + \delta\omega_m$ we get

$$\sum_{n=-N}^{N} (\Gamma + jn\Omega)a_n e^{jn\Omega t} - \frac{j\Delta\omega_m}{2} \sum_{n=-N}^{N} a_n \left[e^{j(n+1)\Omega t} + e^{j(n-1)\Omega t} \right]$$
$$= (-j\mu\sqrt{P_{in}}), \tag{5.62}$$

where $\Gamma = j\Delta\omega + \gamma_r$ and δ_{n0} is the Kronecker delta. By requiring that the coefficients of terms of the same harmonic frequency satisfy the equality in Equation 5.62, we obtain

$$\Gamma a_0 - \frac{j\Delta\omega_m}{2}(a_1 + a_{-1}) = -j\mu\sqrt{P_{in}}, \quad (n = 0), \tag{5.63}$$

$$(\Gamma + jn\Omega)a_n - \frac{j\Delta\omega_m}{2}(a_{n+1} + a_{n-1}) = 0, \quad (n = \pm 1, \pm 2, \ldots \pm(N-1)), \tag{5.64}$$

$$(\Gamma \pm jN\Omega)a_{\mp N} - \frac{j\Delta\omega_m}{2}a_{(\pm N-1)} = 0, \quad (n = \pm N). \tag{5.65}$$

From the above expressions, we can solve for the amplitudes a_n to get

$$a_0 = -j\mu\sqrt{P_{in}}F_0, \tag{5.66}$$

$$a_{\pm n} = \frac{j\Delta\omega_m}{2}F_{\pm n}a_{\pm(n-1)}, \quad (n = 1, 2, \ldots N), \tag{5.67}$$

where the terms $F_{\pm n}$ are computed backward starting from $F_{\pm N}$,

$$F_{\pm N} = \frac{1}{\Gamma \pm jN\Omega}, \tag{5.68}$$

$$F_{\pm n} = \frac{1}{\Gamma \pm jn\Omega + (\Delta\omega_m/2)^2 F_{\pm(n+1)}}, \quad (n = 1, 2, \ldots (N-1)), \tag{5.69}$$

$$F_0 = \frac{1}{\Gamma + (\Delta\omega_m/2)^2(F_1 + F_{-1})}. \tag{5.70}$$

The transmitted power at the output of the modulator is determined from

$$P_{out}(t) = \left|\tilde{s}_{in} - j\mu\tilde{a}\right|^2 = \left|\sqrt{P_{in}} - j\mu\sum_{n=-N}^{N}a_n e^{jn\Omega t}\right|^2. \tag{5.71}$$

Expanding the above expression, we get

$$P_{out}(t) = P_{in} - j\mu\sqrt{P_{in}}\sum_{n=-N}^{N}\left(a_n e^{jn\Omega t} - a_n^* e^{-jn\Omega t}\right) + \mu^2\sum_{n=-N}^{N}\sum_{n'=-N}^{N}a_n a_{n'}^* e^{j(n-n')\Omega t}. \tag{5.72}$$

If we write $P_{out}(t)$ as a sum of harmonic frequencies,

$$P_{out}(t) = \sum_{n=-N}^{N} P_n e^{jn\Omega t},$$

(5.73)

then the power of the nth-order harmonic, P_n, can be obtained by comparing Equations 5.72 and 5.73. Specifically, for the DC component we get

$$P_0 = P_{in} - j\mu\sqrt{P_{in}}(a_0 - a_0^*) + \mu^2 \sum_{n=-N}^{N} |a_n|^2,$$

(5.74)

and for the nth-order harmonic,

$$P_n = -j\mu\sqrt{P_{in}}(a_n - a_{-n}^*) + \mu^2 \sum_{k=-N}^{N} a_k a_{k-n}^*.$$

(5.75)

Note that the expression for the DC component can also be written as

$$P_0 = P_{out}^{DC} + \mu^2 \sum_{n \neq 0} |a_n|^2,$$

(5.76)

where P_{out}^{DC} is the transmitted DC power in the absence of a modulating signal (Equation 5.38), and the sum is over all n terms excluding the DC term. Each of the terms $|a_n|^2$ in the summation represents the contribution to the DC power from the self-beating of each harmonic, which is neglected in the SS analysis. The output AC power of the APMR modulator is given by the first-order harmonic

$$P_1 = -j\mu\sqrt{P_{in}}(a_1 - a_{-1}^*) + \mu^2 \sum_{k=-N}^{N} a_k a_{(k-1)}^*.$$

(5.77)

Higher-order harmonic terms contribute to nonlinear effects. In particular, the most dominant contribution is from the second-order term

$$P_2 = -j\mu\sqrt{P_{in}}(a_2 - a_{-2}^*) + \mu^2 \sum_{k=-N}^{N} a_k a_{(k-2)}^*.$$

(5.78)

However, in certain regimes of operation, the third-order harmonic can also become nearly as large as the second-order harmonic, as shown in the example below.

Figure 5.10a shows the normalized energies $|a_n/A_0|^2$ of the first four harmonics ($n = 0$–3) in an APMR microring as a function of the magnitude of the resonant frequency shift $\delta_m = \Delta\omega_m/\gamma_r$. The energies are normalized with respect to the energy in the microring in the absence of a modulating signal, $|A_0|^2 = |a_0(0)|^2$. The microring has normalized frequency detune $\delta = 1$ and critical coupling (normalized coupling ratio $\gamma = 1$), and the normalized modulation frequency is set at $\tilde{\Omega} = \Omega/\gamma_r = 0.1$. A total of $N = 10$ harmonics are used in the computation. The normalized transmitted powers at the output of the modulator (P_n/P_{in}) of the first, second, and third

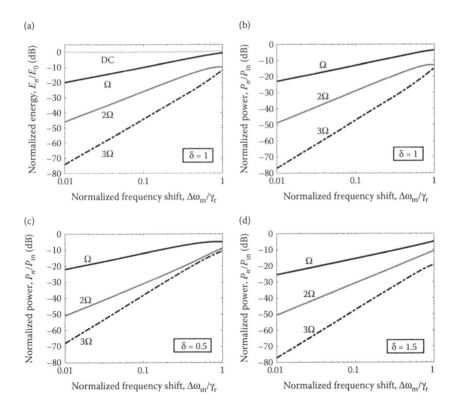

Figure 5.10 (a) Plot of the normalized energies in the microring of the first four harmonics as functions of the frequency shift. Plots of the output modulator powers of the first three harmonics for (b) $\delta = 1$, (c) $\delta = 0.5$, and (d) $\delta = 1.5$. The APMR has critical coupling ($\gamma = 1$).

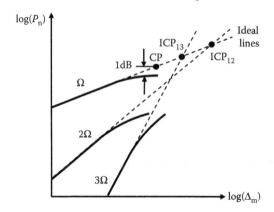

Figure 5.11   Logarithm plot of output harmonic powers versus modulation amplitude showing the 1 dB CP and intercept points ICP_{12}, and ICP_{13}.

harmonics are shown in Figure 5.10b. Figure 5.10c and d show the output harmonics for frequency detune values of $\delta = 0.5$ and 1.5. We observe that the second harmonic dominates for all three cases but the third harmonic can become as strong as the second harmonic at smaller frequency detunes.

There are several parameters that can be used to quantify the degree of nonlinearity of a modulator. One useful parameter is the 1 dB compression point (1 dB CP), which is defined as the modulation amplitude for which the output power of the first-order harmonic falls below the ideal straight line (or the SS response) by 1 dB, as illustrated in Figure 5.11. Another useful parameter is the intercept point, which is used to quantify the significance of the contributions of high-order harmonics to nonlinear distortion. For example, to quantify the relative significance of the second-order and third-order harmonics, we can denote ICP_{12} as the intercept point of the ideal output powers (indicated by straight dashed lines in Figure 5.11) of the first and second harmonics, and ICP_{13} as the intercept point of the first and third harmonics. These parameters are also illustrated in Figure 5.11. In general, the smaller the value of an intercept point, the more significant the contribution of the corresponding harmonic to nonlinear distortion.

Figure 5.12a and b plot the 1 dB CP and the intercept points as functions of the frequency detune δ for an APMR modulator with critical coupling ($\gamma = 1$). The normalized RF modulation frequency is set at $\tilde{\Omega} = 0.1$. We observe in Figure 5.12a that the 1 dB CP value

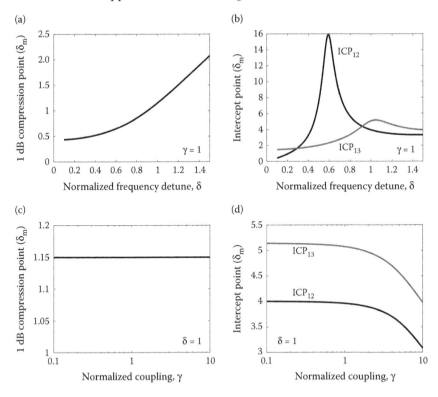

Figure 5.12   Plots of (a) the 1 dB CP and (b) intercept points versus the normalized frequency detune δ for $\gamma = 1$ and $\tilde{\Omega} = 0.1$. Plots of (c) the 1 dB CP and (d) intercept points versus the normalized coupling γ for $\delta = 1$.

increases at larger frequency detunes, which implies that nonlinear distortion is reduced. However, the relative dominance of the second and third harmonics depends strongly on the frequency detune. As Figure 5.12b shows, the ICP13 value is smaller than the ICP12 value over the frequency detune range $\delta \sim 0.3$–0.9 indicating that the third-harmonic component is the dominant source of nonlinearity over this range of frequency detunes. Outside this range, the second-harmonic component always dominates. Figure 5.12c and d show the 1 dB compression points and intercept points as functions of the normalized coupling γ, for a fixed frequency detune $\delta = 1$ and normalized RF frequency $\tilde{\Omega} = 0.1$. We see that the 1 dB CP value is relatively independent of the normalized coupling, and for all γ values, the second harmonic always dominates over the third harmonic.

5.3.3 Intermodulation products

The LS analysis in the previous section can also be extended to the case where the microring is modulated by two RF sinusoidal signals at two slightly different frequencies Ω_1 and Ω_2 (with $\Omega_1 - \Omega_2 \ll \Omega_1, \Omega_2$). Such an analysis is useful for evaluating the second-order and third-order intermodulation products (IMP2 and IMP3), which are defined as the output powers at frequencies $\Omega_1 \pm \Omega_2$ and $2\Omega_1 - \Omega_2$, respectively. The modulating signal is represented as a sum of two sinusoidal waves with frequencies Ω_1 and Ω_2 but the same amplitude $\Delta\omega_m$:

$$\delta\omega_m(t) = \Delta\omega_m\left[\cos(\Omega_1 t) + \cos(\Omega_2 t)\right] = \Delta\omega_m(e^{j\Omega_1 t} + e^{-j\Omega_1 t}$$
$$+ e^{j\Omega_2 t} + e^{-j\Omega_2 t})/2. \tag{5.79}$$

The wave amplitude in the microring has a solution of the form

$$\tilde{a}(t) = \sum_{m=-M}^{M}\sum_{n=-N}^{N} a_{mn}e^{j(\omega + m\Omega_1 + n\Omega_2)t}, \tag{5.80}$$

with a similar expression for the output power $P_{out}(t)$. In Equation 5.80, M and N are the number of harmonics of Ω_1 and Ω_2 on each side of the center frequency ω used in the analysis. By substituting Equations 5.79 and 5.80 into Equation 5.26 and following the procedure for the single-frequency analysis in the previous section, we obtain the solutions for the wave amplitude a_{mn} and the power P_{mn} associated with the harmonic frequency $m\Omega_1 + n\Omega_2$. For a microring modulator, typical intermodulation products of interest are the powers at the sum and difference frequencies $\Omega_1 \pm \Omega_2$, and the third-order intermodulation products $2\Omega_1 - \Omega_2$ and $2\Omega_2 - \Omega_1$. Although the large number of harmonic terms makes the computation somewhat more complex than the single-frequency analysis, the intermodulation product analysis provides a more accurate way to quantify nonlinear distortion in narrowband microring modulators since the intermodulation frequencies are close to the input RF frequencies Ω_1 and Ω_2. Finally, we note that the nonlinear distortions of a microring modulator can be reduced by linearizing its transfer function around the operating point. This approach has been adopted to devise modulators with high linearity based on Mach–Zehnder interferometers loaded with microring resonators (Xie et al. 2003, Van et al. 2006).

5.4 Summary

Active photonic applications of microring resonators typically involve electrically modifying the resonant frequencies of the microrings. In this chapter, we reviewed the various physical mechanisms that can be used to tune or modulate the resonant frequencies of a microring, including the thermo-optic effect, the electro-optic effect, and free carrier dispersion. Using the all-pass microring modulator as a prototypical active microring device, we developed a dynamic model based on energy coupling which allows us to study the device behaviors in both the time and frequency domains and under SS and LS conditions. These analyses also allow us to derive expressions for important parameters characterizing the device performance such as modulation efficiency, electrical bandwidth, harmonic intercept points, and intermodulation products.

References

Agullo-Lopez, F., Cabrera, J. M., Agullo-Rueda, F. 1994. *Electrooptics—Phenomena, Materials and Applications*. London: Academic Press.

Campbell, J. C., Blum, F. A., Shaw, D. W., Lawley, K. L. 1975. GaAs electro-optic directional-coupler switch. *Appl. Phys. Lett.* 27(4): 202–205.

Cocorullo, G., Della Corte, F. G., Rendina, I. 1999. Temperature dependence of the thermo-optic coefficient in crystalline silicon between room temperature and 550 K at the wavelength of 1523 nm. *Appl. Phys. Lett.* 74(22): 3338–3340.

Dong, P., Liao, S., Feng, D., Liang, H., Zheng, D., Shafiiha, R., Kung, C.-C. et al. 2009. Low Vpp, ultralow-energy, compact, high-speed silicon electro-optic modulator. *Opt. Express* 17(25): 22484–22490.

Dong, P., Qian, W., Liang, H., Shafiiha, R., Feng, N., Feng, D., Zheng, X., Krishnamoorthy, A. V., Asghari, M. 2010a. Low power and compact reconfigurable multiplexing devices based on silicon microring resonators. *Opt. Express* 18(10): 9852–9858.

Dong, P., Shafiiha, R., Liao, S., Liang, H., Feng, N., Feng, D., Li, G., Zheng, X., Krishnamoorthy, A. V., Asghari, M. 2010b. Wavelength-tunable silicon microring modulator. *Opt. Express* 18(11): 10941–10946.

Leviton, D. B., Frey, B. J. 2006. Temperature-dependent absolute refractive index measurements of synthetic fused silica. In *SPIE Astronomical Telescopes + Instrumentation*, Orlando, FL, 62732K.

Li, C., Zhou, L., Poon, A. W. 2007. Silicon microring carrier-injection-based modulators/switches with tunable extinction ratios and OR-logic switching by using waveguide cross-coupling. *Opt. Express* 15(8): 5069–5076.

Liu, A., Jones, R., Liao, L., Samara-Rubio, D., Rubin, D., Cohen, O., Nicolaescu, R., Paniccia, M. 2004. A high-speed silicon optical modulator based on a metal-oxide-semiconductor capacitor. *Nature* 427(12): 615–618.

Milivojevic, B., Raabe, C., Shastri, A., Webster, M., Metz, P., Sunder, S., Chattin, B., Wiese, S., Dama, B., Shastri, K. 2013. 112Gb/s DP-QPSK transmission over 2427 km SSMF using small-size silicon photonic IQ modulator and low-power CMOS driver. In *Optical Fiber Communication Conference*, Anaheim, CA, paper OTh1D.1.

Oh, M.-C., Zhang, H., Szep, A., Chuyanov, V., Steier, W. H., Zhang, C., Dalton, L. R., Erlig, H., Tsap, B., Fetterman, H. R. 2000. Electro-optic polymer modulators for 1.55 µm wavelength using phenyltetrane bridged chromophore in polycarbonate. *Appl. Phys. Lett.* 76(24): 3525–3527.

Pile, B., Taylor, G. 2014. Small-signal analysis of microring resonator modulators. *Opt. Express* 22(12): 14913–14928.

Rabiei, P., Steier, W. H., Zhang, C., Dalton, L. R. 2002. Polymer micro-ring filters and modulators. *J. Lightwave Technol.* 20(11): 1968–1975.

Raghunathan, V., Ye, W. N., Hu, J., Izuhara, T., Michel, J., Kimerling, L. 2010. Athermal operation of silicon waveguides: spectral, second order and footprint dependencies. *Opt. Express* 18(7): 17631–17639.

Sacher, W. D., Green, W. M. J., Assefa, S., Barwicz, T., Pan, H., Shank, S. M., Vlasov, Y. A., Poon, J. S. K. 2013. Coupling modulation of microrings at rates beyond the linewidth limit. *Opt. Express* 21(8): 9722–9733.

Sacher, W. D., Poon, J. K. S. 2008. Dynamics of microring resonator modulators. *Opt. Express* 16(20): 15741–15753.

Sherwood-Droz, N., Wang, H., Chen, L., Lee, B. G., Biberman, A., Bergman, K., Lipson, M. 2008. Optical 4x4 hitless silicon router for optical Networks-on-Chip (NoC). *Opt. Express* 16(20): 15915–15922.

Sze, S. M. 1981. *Physics of Semiconductor Devices*, 2nd Ed. New York: John Wiley & Sons.

Van, V., Herman, W. N., Ho, P.-T. 2006. Linearized microring-loaded Mach-Zehnder modulator with RF gain. *J. Lightwave Technol.* 24(4): 1850–1854.

Xie, X., Khurgin, J., Kang, J., Chow, F.-S. 2003. Linearized Mach–Zehnder intensity modulator. *IEEE Photonics Technol. Lett.* 15(4): 531–533.

Xu, Q., Schmidt, B., Pradhan, S., Lipson, M. 2005. Micrometre-scale silicon electro-optic modulator. *Nature* 435(19): 325–327.

Yariv, A. 1991. *Optical Electronics*, 4th Ed. Philadelphia: Saunders College Publishing.

Index